Contributions in Algebra and Algebraic Geometry

CONTEMPORARY MATHEMATICS

738

Contributions in Algebra and Algebraic Geometry

International Conference
Algebra, Discrete Mathematics and Applications
December 9–11, 2017
Dr. Babasaheb Ambedkar Marathwada University
Aurangabad (Maharashtra), India

Shrikrishna G. Dani
Surender K. Jain
Jugal K. Verma
Meenakshi P. Wasadikar
Editors

AMS AMERICAN MATHEMATICAL SOCIETY

2010 *Mathematics Subject Classification.* Primary 13-xx, 14-xx, 15-xx, 16-xx.

Library of Congress Cataloging-in-Publication Data

Names: International Conference on Algebra, Discrete Mathematics and Applications (2017 : Aurangabad, India) | Dani, S. G., editor. | Jain, S. K. (Surender Kumar), 1938– editor. | Verma, Jugal, editor. | Wasadikar, Meenakshi P., editor.

Title: Contributions in algebra and algebraic geometry : International Conference on Algebra, Discrete Mathematics and Applications, December 9–11, 2017, Dr. Babasaheb Ambedkar Marathwada University, Aurangabad (Maharashtra), India / Shrikrishna G. Dani, Surender K. Jain, Jugal K. Verma, Meenakshi P. Wasadikar, editors.

Description: Providence, Rhode Island : American Mathematical Society, [2019] | Series: Contemporary mathematics ; volume 738 | Includes bibliographical references.

Identifiers: LCCN 2019019104 | ISBN 9781470447359 (alk. paper)

Subjects: LCSH: Algebra–Congresses. | Geometry, Algebraic–Congresses. | AMS: Commutative algebra. msc | Algebraic geometry. msc | Linear and multilinear algebra; matrix theory. msc | Associative rings and algebras. msc

Classification: LCC QA150 .I5655 2017 | DDC 512–dc23

LC record available at https://lccn.loc.gov/2019019104

Contemporary Mathematics ISSN: 0271-4132 (print); ISSN: 1098-3627 (online)

DOI: https://doi.org/10.1090/conm/738

∞ The paper used in this book is acid-free and falls within the guidelines
established to ensure permanence and durability.
Visit the AMS home page at `https://www.ams.org/`

10 9 8 7 6 5 4 3 2 1   24 23 22 21 20 19

Contents

Preface viii

Leavitt path algebras over a poset of fields
PERE ARA 1

Roots of elements in Lie groups and the exponential maps
S. G. DANI 13

A note on Nullstellensatz over finite fields
SUDHIR R. GHORPADE 23

Minkowski inequality and equality for multiplicity of ideals of finite length in Noetherian local rings
KRITI GOEL, R. V. GURJAR, and J. K. VERMA 33

Decomposition of singular elements of an algebra into product of idempotents, a survey
S. K. JAIN and A. LEROY 57

Some results on integrally closed domains
ANUJ JAKHAR, SUDESH K. KHANDUJA, and NEERAJ SANGWAN 75

On examples of Baer and Rickart module hulls
JAE KEOL PARK and S. TARIQ RIZVI 81

Tableau correspondences and representation theory
DIGJOY PAUL, AMRITANSHU PRASAD, and ARGHYA SADHUKHAN 109

A relationship between Gelfand-Tsetlin bases and Chari-Loktev bases for irreducible finite dimensional representations of special linear Lie algebras
K. N. RAGHAVAN, B. RAVINDER, and SANKARAN VISWANATH 125

Algebra of generic matrices is not coherent
LANCE SMALL and EFIM ZELMANOV 135

Partial endomorphisms of almost self injective modules
SURJEET SINGH 139

Preface

This volume consists of texts of invited lectures in the area of algebra and algebraic geometry delivered at the International Conference on Algebra, Discrete Mathematics and Applications. The conference was organized by Professor Meenakshi Wasadikar in cooperation with Professor Shriram Nimbhorkar at the Department of Mathematics, Dr. Babasaheb Ambedkar Marathwada University, Aurangabad, India, during December 9–11, 2017, in honor of Fields Medalist Professor Efim Zelmanov.

The invited speakers included Professors Pere Ara (Universitat Autonoma de Barcelona, Spain), Arvind Ayyer (IISc, Bangalore, India), S. G. Dani (UM-DAE Centre for Excellence in Basic Sciences, University of Mumbai, Mumbai, India), A. Facchini (Padova, Italy), S. R. Ghorpade (IIT Bombay, Mumbai, India), S. K. Jain (Ohio, USA), S. K. Khanduja (IISER, Mohali, India), Andre Leroy (Lens, France), Amritanshu Prasad (IMSc., Chennai), K. N. Raghavan (IMSc., Chennai), S. T. Rizvi (Ohio, USA), Surjeet Singh (San Francisco, USA), Jugal K. Verma (IIT Bombay, Mumbai, India), B. N. Waphare (Savitribai Phule Pune University, Pune, India), and Efim Zelmanov (University of San Diego, CA, USA). Apart from the invited lectures, there were fifty contributory talks. More than 150 delegates participated in the conference.

All the papers in the volume are on topics in the mainstream of research in the field of algebra and algebraic geometry. The papers were refereed to uphold high standards. An important feature of this volume is that it contains some expository articles that can inspire interested researchers in the field to work on the questions raised in the articles.

The conference was inaugurated by Professor N. K. Thakare (Founder Vice-Chancellor, North Maharashtra University, Jalgaon, India) and presided over by Professor B. A. Chopade, Vice-Chancellor, Dr. Babasaheb Ambedkar Marathwada University, Aurangabad, India. During the inauguration, Professors Efim Zelmanov, S. K. Jain, S. B. Nimse (Ex-Vice-Chancellor, Lucknow University, Lucknow and Swami Ramanand Tirth Marathwada University, Nanded, India) spoke on the positive outcomes of organizing such conferences. Among other benefits these conferences provide inspiration to the younger members of the community.

The conference was supported by the Commission for Developing Countries (CDC) of the International Mathematical Union (IMU), Berlin, the National Board for Higher Mathematics (NBHM) of the Department of Atomic Energy, Mumbai, the Science and Engineering Research Board (SERB) of the Department of Science and Technology, New Delhi and the Defense Research and Development Organization (DRDO), New Delhi.

This volume owes its genesis to the conference mentioned above and it would be appropriate to record here thanks to all those who contributed to its organization and success. We are thankful to Professors B. A. Chopade (Vice-Chancellor), Sadhana Pande (Registrar), W. K. Sarwade (Director, BCUD) and other concerned authorities of Dr. Babasaheb Ambedkar Marathwada University, Aurangabad, for helping us in the organization of the conference. We thank Professors N. K. Thakare, Ajit I. Singh and Shriram Nimbhorkar for various suggestions regarding organization and funding. The whole-hearted support extended by Principal Madhukar Pawar and students Uttara, Karuna, Vidya, Yogita, Amit, Deepali and Jaya will always be a sweet memory. Thanks to each of them.

We take this opportunity to thank all the referees who gave their valuable time to referee the papers and who honored the deadlines given to them.

Last but not least, we profusely thank Christine M. Thivierge of the American Mathematical Society. She worked so diligently throughout the process, from the beginning until the volume came to fruition, promptly answering any question that we raised and sending timely reminders about deadlines.

Editors:

Shrikrishna G. Dani,

Surender K. Jain,

Jugal K. Verma, and

Meenakshi P. Wasadikar.

April 15, 2019

Contemporary Mathematics
Volume **738**, 2019
https://doi.org/10.1090/conm/738/14874

Leavitt path algebras over a poset of fields

Pere Ara

ABSTRACT. Let E be a finite directed graph, and let I be the poset defined by the antisymmetrization of its set of vertices E^0 with respect to a pre-order $\le$ satisfying that $v \le w$ whenever there exists a directed path from w to v, for $v, w \in E^0$. Assuming that I is a tree, we define a poset of fields over I as a family $\mathbf{K} = \{K_i : i \in I\}$ of fields K_i such that $K_i \subseteq K_j$ if $j \le i$. We define the concepts of a Leavitt path algebra $L_{\mathbf{K}}(E)$ and a regular algebra $Q_{\mathbf{K}}(E)$ over the poset of fields $\mathbf{K}$, and we show that $Q_{\mathbf{K}}(E)$ is a hereditary von Neumann regular ring, and that its monoid $\mathcal{V}(Q_{\mathbf{K}}(E))$ of isomorphism classes of finitely generated projective modules is canonically isomorphic to the graph monoid $M(E)$ of E.

Introduction

The work in the present paper is instrumental for the constructions developed in [**4**] (see also [**5**]). The main result in [**4**] states that for every finitely generated conical refinement monoid M and every field K there exists a von Neumann regular K-algebra R such that $\mathcal{V}(R) \cong M$. The algebra R is a certain universal localization of a precisely defined *Steinberg algebra*. The class of Steinberg algebras has been intensively studied in the last few years, see for instance the recent survey [**17**] and its references. It includes as prominent examples Leavitt path algebras of arbitrary graphs. For the proof of the main result of [**4**], one needs to interpret the algebra R as an algebra obtained using certain building blocks, using constructions which were partially introduced in [**3**] and fully developed in [**4**]. The purpose of this paper is to introduce these building blocks in a general way, and to show their basic properties. These algebras generalize the main construction in [**6**] of the regular algebra $Q_K(E)$ of a finite directed graph E. The main features of the K-algebra $Q_K(E)$ are that it is von Neumann regular and that $\mathcal{V}(Q_K(E)) \cong M(E)$, where $M(E)$ is the *graph monoid* of E (see Section 2 for the definition of the graph monoid).

This is generalized here as follows. Let E be a finite graph and let $\le$ be a pre-order on I such that $v \le w$ whenever there exists a directed path from w to v. Let $I = E^0/{\sim}$ be the antisymmetrization of E^0 with respect to the pre-order $\le$, endowed with its canonical partial order, also denoted by $\le$. We assume throughout

2010 *Mathematics Subject Classification.* Primary 16D70; Secondary 06A12, 06F05, 46L80.
Key words and phrases. von Neumann regular ring, path algebra, Leavitt path algebra, universal localization.
Partially supported by DGI-MINECO-FEDER through the grant MTM2017-83487-P and by the Generalitat de Catalunya through the grant 2017-SGR-1725.

that I is a *tree*, that is, that there exists a maximum element i_0 in I and that for each $i \in I$ the interval $[i, i_0]$ is a chain. Under these assumptions, we define a poset of fields $\mathbf{K} = \{K_i : i \in I\}$ as a collection of fields K_i, for $i \in I$, such that $K_i \subseteq K_j$ whenever $j \leq i$. The usual Leavitt path algebra $L_K(E)$ and regular algebra $Q_K(E)$ over a field K are here generalized to a Leavitt path algebra $L_{\mathbf{K}}(E)$ and a regular algebra $Q_{\mathbf{K}}(E)$ with coefficients in the given poset of fields $\mathbf{K}$. The main results state that $Q_{\mathbf{K}}(E)$ is a hereditary von Neumann regular ring and that $\mathcal{V}(Q_{\mathbf{K}}(E)) \cong M(E)$ canonically. Our techniques here extend the ones introduced in [**7**], where a more restricted scope was used to study algebras over an ordered finite sequence of fields.

For more information on the realization problem we refer the reader to [**6**], [**3**], [**8**], and to the survey papers [**15**], [**2**].

1. Preliminary definitions

All rings in this paper will be associative and all monoids will be abelian. A (not necessarily unital) ring R is *von Neumann regular* if for every $a \in R$ there is $b \in R$ such that $a = aba$. Our basic reference for the theory of von Neumann regular rings is [**14**].

For a unital ring R, the monoid $\mathcal{V}(R)$ is the monoid of isomorphism classes of finitely generated projective left R-modules, where the operation is induced by direct sum. If R is an exchange ring (in particular, if R is von Neumann regular), then $\mathcal{V}(R)$ is a conical refinement monoid, see [**9**, Corollary 1.3].

Say that a subset A of a poset I is a *lower subset* in case $q \leq p$ and $p \in A$ imply $q \in A$. We use $I \downarrow i$ to denote the lower subset of I consisting of all the elements j such that $j \leq i$.

For an element p of a poset I, write

$$\mathrm{L}(p) = \mathrm{L}(I, p) = \{q \in I : q < p \text{ and } [q, p] = \{q, p\}\}.$$

An element of $\mathrm{L}(p)$ is called a *lower cover* of p.

In the following, K will denote a field and $E = (E^0, E^1, r, s)$ a finite quiver (directed graph). Here $s(e)$ is the *source vertex* of the arrow e, and $r(e)$ is the *range vertex* of e. A *path* in E is either an ordered sequence of arrows $\alpha = e_1 \cdots e_n$ with $r(e_t) = s(e_{t+1})$ for $1 \leq t < n$, or a path of length 0 corresponding to a vertex $v \in E^0$. The paths of length 0 are called trivial paths. A non-trivial path $\alpha = e_1 \cdots e_n$ has length n and we define $s(\alpha) = s(e_1)$ and $r(\alpha) = r(e_n)$. We will denote the length of a path α by $|\alpha|$, the set of all paths of length n by E^n, for $n > 1$, and the set of all paths by $\mathrm{Path}(E)$.

For $v, w \in E^0$, set $v \geq w$ in case there is a (directed) path from v to w. A subset H of E^0 is called *hereditary* if $v \geq w$ and $v \in H$ imply $w \in H$.

Let us recall the construction from [**6**] of the regular algebra $Q_K(E)$ of a quiver E, although we will follow the presentation in [**3**] rather than the used in [**6**]. That is, relations (CK1) and (CK2) below are reversed with respect to their counterparts in [**6**], so that we are led to work primarily with *left* modules instead of right modules. Therefore we recall the basic features of the regular algebra $Q_K(E)$ in terms of the notation used here. We will only need finite quivers in the present paper, so we restrict attention to them. The algebra $Q_K(E)$ fits into the following commutative diagram of injective algebra morphisms:

$$\begin{array}{ccccccc} K^{E^0} & \longrightarrow & P_K(E) & \xrightarrow{\iota_\Sigma} & P_K^{\mathrm{rat}}(E) & \longrightarrow & P_K((E)) \\ \downarrow & & \iota_{\Sigma_1}\downarrow & & \iota_{\Sigma_1}\downarrow & & \iota_{\Sigma_1}\downarrow \\ P_K(E^*) & \longrightarrow & L_K(E) & \xrightarrow{\iota_\Sigma} & Q_K(E) & \longrightarrow & U_K(E). \end{array}$$

Here $P_K(E)$ is the path K-algebra of E, E^* denotes the inverse quiver of E, that is, the quiver obtained by reversing the orientation of all the arrows in E, $P_K((E))$ is the algebra of formal power series on E, and $P_K^{\mathrm{rat}}(E)$ is the algebra of rational series, which is by definition the division closure of $P_K(E)$ in $P_K((E))$ (which agrees with the rational closure [**6**, Observation 1.18]). The maps ι_Σ and ι_{Σ_1} indicate universal localizations with respect to the sets Σ and Σ_1 respectively. (We refer the reader to [**13**] and [**18**] for the general theory of universal localization.) Here Σ is the set of all square matrices over $P_K(E)$ that are sent to invertible matrices by the augmentation map $\epsilon\colon P_K(E) \to K^{E^0}$. By [**6**, Theorem 1.20], the algebra $P^{\mathrm{rat}}(E)$ coincides with the universal localization $P_K(E)\Sigma^{-1}$. The set $\Sigma_1 = \{\mu_v \mid v \in E^0,\ s^{-1}(v) \neq \emptyset\}$ is the set of morphisms between finitely generated projective left $P_K(E)$-modules defined by

$$\begin{aligned} \mu_v\colon P_K(E)v &\longrightarrow \bigoplus_{i=1}^{n_v} P_K(E)r(e_i^v) \\ r &\longmapsto \left(re_1^v, \dots, re_{n_v}^v\right) \end{aligned}$$

for any $v \in E^0$ such that $s^{-1}(v) \neq \emptyset$, where $s^{-1}(v) = \{e_1^v, \dots, e_{n_v}^v\}$. By a slight abuse of notation, we use also μ_v to denote the corresponding maps between finitely generated projective left $P_K^{\mathrm{rat}}(E)$-modules and $P_K((E))$-modules respectively.

The following relations hold in $Q_K(E)$:

(V) $vv' = \delta_{v,v'}v$ for all $v, v' \in E^0$.
(E1) $s(e)e = er(e) = e$ for all $e \in E^1$.
(E2) $r(e)e^* = e^*s(e) = e^*$ for all $e \in E^1$.
(CK1) $e^*e' = \delta_{e,e'}r(e)$ for all $e, e' \in E^1$.
(CK2) $v = \sum_{\{e \in E^1 \mid s(e)=v\}} ee^*$ for every $v \in E^0$ that emits edges.

The Leavitt path algebra $L_K(E) = P_K(E)\Sigma_1^{-1}$ is the algebra generated by $\{v \mid v \in E^0\} \cup \{e, e^* \mid e \in E^1\}$ subject to the relations (V)–(CK2) above, see [**1**]. By [**6**, Theorem 4.2], the algebra $Q_K(E)$ is a von Neumann regular hereditary ring and $Q_K(E) = P_K(E)(\Sigma \cup \Sigma_1)^{-1}$. Here the set Σ can be clearly replaced with the set of all square matrices of the form $I_n + B$ with $B \in M_n(P_K(E))$ satisfying $\epsilon(B) = 0$, for all $n \geq 1$.

2. The results

Our graphs are shaped by a poset with a special property, as follows:

DEFINITION 2.1. Let $(I, \leq)$ be a poset. We say that I is a *tree* in case there is a maximum element $i_0 \in I$ and for every $i \in I$ the interval $[i, i_0] := \{j \in I \mid i \leq j \leq i_0\}$ is a chain. The element i_0 will be called the *root* of the tree I.

Let $\leq$ be a preorder on E^0 such that $w \geq v$ whenever there exists a path in E from w to v. Let I be the antisymmetrization of E^0, with the partial order $\leq$ induced by the order on E^0. Thus, denoting by $[v]$ the class of $v \in E^0$ in I, we have $[v] \leq [w]$ if and only if $v \leq w$.

We will assume throughout this section that E is a finite quiver such that $I := E/\sim$ is a tree.

For $v \in E^0$, we refer to the set $[v]$ as the *component* of v, and we will denote by $E[v]$ the restriction of E to $[v]$, that is, the graph with $E[v]^0 = [v]$ and $E[v]^1 = \{e \in E^1 \mid s(e) \in [v] \text{ and } r(e) \in [v]\}$.

We will assume henceforth that we are given a family $\mathbf{K} = \{K_i\}_{i\in I}$ of fields such that $K_j \subseteq K_i$ if $i \leq j$. We refer to this family as a *poset of fields* (over I). We will define a certain K_0-algebra $Q_{\mathbf{K}}(E)$, where $K_0 := K_{[v_0]}$.

Let $i, j \in I$, and suppose that $k \in (I\downarrow i) \cap (I \downarrow j)$. Then, since $[k, i_0]$ is a chain, we must have $i \leq j$ or $j \leq i$. Therefore, if i and j are incomparable elements of I then $(I \downarrow i) \cap (I \downarrow j) = \emptyset$. It follows that if J is a lower subset of I, and $x_1, \dots, x_t$ are the maximal elements of J, then $J = \bigsqcup_{i=1}^t (I \downarrow x_i)$.

Given a lower subset J of I we consider the set of vertices

$$E^0_J = \{v \in E^0 \mid [v] \in J\}.$$

Observe that E^0_J is a hereditary subset of E^0. We will denote by E_J the restriction graph $E|_{E^0_J}$ corresponding to the hereditary subset E^0_J of E^0, and we set $p_J := \sum_{v\in E^0_J} v$.

We retain the above notation. We first consider an algebra of power series $P_{\mathbf{K}}((E))$. The algebra $P_{\mathbf{K}}((E))$ is defined as the algebra of formal power series of the form $a = \sum_{\gamma\in \mathrm{Path}(E)} a_\gamma \gamma$, where each $a_\gamma \in K_{[r(\gamma)]}$. The usual multiplication of formal power series gives an structure of algebra over K_0 on $P_{\mathbf{K}}((E))$. Observe that if $a_\gamma \in K_{[r(\gamma)]}$, $b_\mu \in K_{[r(\mu)]}$, and $r(\gamma) = s(\mu)$, then since $[r(\gamma)] \geq [r(\mu)]$, we have that

$$a_\gamma b_\mu \in K_{[r(\mu)]} = K_{[r(\gamma\mu)]},$$

because $K_{[r(\gamma)]} \subseteq K_{[r(\mu)]}$.

The path algebra $P_{\mathbf{K}}(E)$ is defined as the subalgebra of $P_{\mathbf{K}}((E))$ consisting of all the series in $P_{\mathbf{K}}((E))$ having finite support.

For $i \in I$, let $\mathbf{K}_i$ be the system of fields defined over $I \downarrow i$ by $\mathbf{K}_i = \{K_j\}_{j\leq i}$. Then one may define the algebras $P_{\mathbf{K}_i}((E_{I\downarrow i}))$ and $P_{\mathbf{K}_i}(E_{I\downarrow i})$. Note that for $i \in I$ we have

$$P_{\mathbf{K}_i}(E_{I\downarrow i}) = P_{K_i}(E_{I\downarrow i}) + \Big(\bigoplus_{j\in \mathrm{L}(I,i)} P_{K_i}(E_{I\downarrow i}) P_{\mathbf{K}_j}(E_{I\downarrow j})\Big).$$

We consider the commutative algebra $\mathcal{E} := \bigoplus_{i\in I} \bigoplus_{v\in i} K_i v$, which is a subalgebra of $P_{\mathbf{K}}(E)$. Note that there is an augmentation map $\epsilon\colon P_{\mathbf{K}}((E)) \to \mathcal{E}$, and that a square matrix $A \in M_n(P_{\mathbf{K}}((E)))$ is invertible if and only if $\epsilon(A)$ is invertible in $M_n(\mathcal{E})$ (see e.g. the proof of [**6**, Lemma 1.8]).

Recall from Section 1 the definition and basic properties of the algebra $P_K^{\mathrm{rat}}(E)$ of rational series, for a field K. We are now ready to define the algebra of rational series in our setting. Let E and $\mathbf{K} = \{K_i\}_{i\in I}$ be as above. We will define inductively $P_J^{\mathrm{rat}}(E)$ for any lower subset J of I.

We first set

$$P_S^{\mathrm{rat}}(E) = \bigoplus_{i\in S} P_{K_i}^{\mathrm{rat}}(E_{\{i\}}),$$

for any non-empty subset S of minimal elements of I. Now assume that J is a non-empty lower subset of E and that we have defined the algebras $P_{J''}^{\mathrm{rat}}(E)$ for all lower subsets J'' of J. Assuming that $J \neq E^0$, let i be a minimal element in $E^0 \setminus J$.

We now define $P^{\mathrm{rat}}_{J'}(E)$, where $J' = J \cup \{i\}$, as

$$P^{\mathrm{rat}}_{J'}(E) = P^{\mathrm{rat}}_{K_i}(E_{I\downarrow i}) + P^{\mathrm{rat}}_{K_i}(E_{I\downarrow i})P^{\mathrm{rat}}_{J}(E) + P^{\mathrm{rat}}_{J}(E).$$

This defines inductively (and unambiguously) $P^{\mathrm{rat}}_{J}(E)$ for any non-empty lower subset J of I. Note that if $i_1, \dots, i_r$ are the maximal elements of J then

$$P^{\mathrm{rat}}_{J}(E) = \bigoplus_{k=1}^{r} P^{\mathrm{rat}}_{I\downarrow i_k}(E).$$

We now define $P^{\mathrm{rat}}_{\mathbf{K}}(E) := P^{\mathrm{rat}}_{E^0}(E)$. Observe that $P^{\mathrm{rat}}_{\mathbf{K}}(E)$ is a K_0-subalgebra of $P_{\mathbf{K}}((E))$.

The following generalizes [**6**, Theorem 1.20].

THEOREM 2.2. Let E and $\mathbf{K} = \{K_i\}_{i\in I}$ be as above. Let Σ denote the set of matrices over $P_{\mathbf{K}}(E)$ that are sent to invertible matrices by ϵ. Then $P^{\mathrm{rat}}_{\mathbf{K}}(E)$ is the rational closure of $P_{\mathbf{K}}(E)$ in $P_{\mathbf{K}}((E))$, and the natural map $P_{\mathbf{K}}(E)\Sigma^{-1} \to P^{\mathrm{rat}}_{\mathbf{K}}(E)$ is an isomorphism.

PROOF. We will prove by induction that $P^{\mathrm{rat}}_{I\downarrow i}(E)$ is the rational closure of $P_{\mathbf{K}_i}(E_i)$ in $P_{\mathbf{K}_i}((E_i))$, where $E_i := E_{I\downarrow i}$ is the restriction graph of E to $E^0_{I\downarrow i}$. Let $i_1, \dots, i_r$ be the set of maximal elements of $(I \downarrow i) \setminus \{i\}$, and assume that the result is known for the graphs $E_{i_k} = E_{I\downarrow i_k}$, for $k = 1, \dots, r$. Changing notation we may furthermore assume that $i = i_0$ is the maximal element of I.

Therefore, we have to show that

$$S := P^{\mathrm{rat}}_{\mathbf{K}}(E) = P^{\mathrm{rat}}_{K_0}(E) + P^{\mathrm{rat}}_{K_0}(E)\Big(\bigoplus_{k=1}^{r} P^{\mathrm{rat}}_{\mathbf{K}_{i_k}}(E_{i_k})\Big)$$

is the rational closure of $R := P_{\mathbf{K}}(E)$ in $P_{\mathbf{K}}((E))$. Write $\mathcal{R}$ for this rational closure, and recall that we are assuming that each $P^{\mathrm{rat}}_{\mathbf{K}_{i_k}}(E_{i_k})$ is the rational closure of $P_{\mathbf{K}_{i_k}}(E_{i_k})$ in $P_{\mathbf{K}_{i_k}}((E_{i_k}))$.

It is convenient to introduce some additional notation. Let H be the hereditary subset of E^0 generated by $i_1, \dots, i_k$, i.e., $H = \bigsqcup_{k=1}^{r} E^0_{I\downarrow i_k}$. Recall also the notation $p_H = \sum_{v\in H} v$. Also, to somewhat simplify the notation we set $K := K_0$.

We start by showing that $S \subseteq \mathcal{R}$. Since $P^{\mathrm{rat}}_{K}(E)$ is the rational closure of $P_K(E)$ inside $P_K((E))$ ([**6**, Theorem 1.20]), we see that $P^{\mathrm{rat}}_{K}(E) \subseteq \mathcal{R}$. Also, note that the algebra $p_H\mathcal{R} = p_H\mathcal{R}p_H$ is rationally closed in $p_H P_{\mathbf{K}_H}((E_H))$ and contains $p_H P_{\mathbf{K}_H}(E_H)$, so it must contain the rational closure of $p_H P_{\mathbf{K}_H}(E_H)$ in $p_H P_{\mathbf{K}_H}((E_H))$ which is $p_H P^{\mathrm{rat}}_{\mathbf{K}_H}(E_H)$ by the induction hypothesis. It follows that $P^{\mathrm{rat}}_{K}(E)$ and $p_H P^{\mathrm{rat}}_{\mathbf{K}_H}(E_H)$ are both contained in $\mathcal{R}$. Since $\mathcal{R}$ is a ring, we get $S \subseteq \mathcal{R}$.

To show the reverse inclusion $\mathcal{R} \subseteq S$, take any element a in $\mathcal{R}$. By [**13**, Theorem 7.1.2], there exist a row $\lambda \in {}^nR$, a column $\rho \in R^n$ and a matrix $B \in M_n(R)$ with $\epsilon(B) = 0$ such that

$$a = \lambda(I - B)^{-1}\rho. \tag{2.1}$$

Now the matrix B can be written as $B = B_1 + B_2$, where $B_1 \in M_n(P_{K_0}(E)) \subseteq M_n(R)$ and $B_2 \in M_n(R)$ satisfy that $\epsilon(B_1) = \epsilon(B_2) = 0$, all the entries of B_1 are supported on paths ending in $E^0 \setminus H = [v_0]$ and all the entries of B_2 are supported on paths ending in H. Note that, since H is hereditary, this implies that all the paths in the support of the entries of B_1 start in $E^0 \setminus H$ and thus $B_2B_1 = 0$. It follows that

$$(I - B)^{-1} = (I - B_1 - B_2)^{-1} = (I - B_1)^{-1}(I - B_2)^{-1}, \tag{2.2}$$

and therefore $(I-B)^{-1} = (I-B_1)^{-1} + (I-B_1)^{-1}B_2(I-B_2)^{-1} \in M_n(S)$. It follows from (2.1) that $a \in S$, as desired.

Since the set Σ is precisely the set of square matrices over R which are invertible over $P_{\mathbf{K}}((E))$, we get from a well-known general result (see for instance [**16**, Lemma 10.35(3)]) that there is a surjective K-algebra homomorphism $\phi\colon R\Sigma^{-1} \to \mathcal{R}$.

The rest of the proof is devoted to show that ϕ is injective. We have a commutative diagram

$$\begin{array}{ccc} P_K(E)\Sigma(\epsilon_K)^{-1} & \longrightarrow & R\Sigma^{-1} \\ \phi_K \big\downarrow \cong & & \phi \big\downarrow \\ P_K^{\mathrm{rat}}(E) & \longrightarrow & \mathcal{R} \end{array} \tag{2.3}$$

where the map ϕ_K is an isomorphism by [**6**, Theorem 1.20]. The map $P_K(E)\Sigma(\epsilon_K)^{-1} \to \mathcal{R}$ is injective, so the map $P_K(E)\Sigma(\epsilon_K)^{-1} \to R\Sigma^{-1}$ must also be injective. Hence the K-subalgebra of $R\Sigma^{-1}$ generated by $P_K(E)$ and the entries of the inverses of matrices in $\Sigma(\epsilon_K)$ is isomorphic to $P_K^{\mathrm{rat}}(E)$. Observe that we can replace Σ by the set of matrices of the form $I-B$, where B is a square matrix over R with $\epsilon(B) = 0$. An element x in $R\Sigma^{-1}$ is of the form

$$x = \lambda(I - B)^{-1}\rho \tag{2.4}$$

with $\lambda \in {}^nR$ and $\rho \in R^n$, and $B \in M_n(R)$ satisfies $\epsilon(B) = 0$.

Claim 1. We have

$$p_H R\Sigma^{-1} \cong p_H P_{\mathbf{K}_H}^{\mathrm{rat}}(E_H) \cong P_{\mathbf{K}_H}(E_H)\Sigma(\epsilon_{\mathbf{K}_H})^{-1}.$$

Proof of Claim 1. Observe first that we have a natural homomorphism

$$\psi\colon P_{\mathbf{K}_H}(E_H)\Sigma(\epsilon_{\mathbf{K}_H})^{-1} \longrightarrow p_H R\Sigma^{-1}.$$

By the induction hypothesis, the composition of ψ with the map ϕ is injective, because it coincides with the canonical map from $P_{\mathbf{K}_H}(E_H)\Sigma(\epsilon_{\mathbf{K}_H})^{-1}$ onto $p_H P_{\mathbf{K}_H}^{\mathrm{rat}}(E_H) \cong P_{\mathbf{K}_H}^{\mathrm{rat}}(E_H)$. Hence ϕ induces an isomorphism from

$$\mathfrak{S} := \psi(P_{\mathbf{K}_H}(E_H)\Sigma(\epsilon_{\mathbf{K}_H}^H)^{-1})$$

onto $p_H P_{\mathbf{K}_H}^{\mathrm{rat}}(E_H)$. Note that $\mathfrak{S}$ is precisely the subalgebra of $p_H R\Sigma^{-1}$ generated by $p_H P_{\mathbf{K}_H}(E_H)$ and the entries of the inverses of matrices of the form $p_H I - B$, with B a square matrix over $p_H P_{\mathbf{K}_H}(E_H)$ with $\epsilon(B) = 0$. For an element x in $R\Sigma^{-1}$, we write it in its canonical form (2.4) and we write $B = B_1 + B_2$ with all the entries in B_1 ending in $E^0 \setminus H$ and all the entries of B_2 ending in H.

Now multiply (2.4) on the left by p_H and use (2.2) to get

$$\begin{aligned} p_H x &= p_H\lambda(I - B_1)^{-1}\rho + p_H\lambda(I - B_1)^{-1}B_2(I - B_2)^{-1}\rho \\ &= p_H\lambda p_H(I - B_1)^{-1}\rho + p_H\lambda p_H(I - B_1)^{-1}B_2(I - B_2)^{-1}\rho \\ &= p_H\lambda p_H\rho + p_H\lambda p_H B_2 p_H(I - B_2)^{-1}\rho. \end{aligned}$$

Write $B_2 = B_2' + B_2''$, where all the entries of B_2' start in $E^0 \setminus H$ and all the entries in B_2'' start in H (and so end in H as well). Note that $(I - B_2')^{-1} = I + B_2'$, because $B_2'^2 = 0$, so that $p_H(I-B_2')^{-1} = p_H$. Since $B_2''B_2' = 0$ we have $(I-B_2)^{-1} = (I-B_2')^{-1}(I-B_2'')^{-1}$, and thus

$$p_H x = p_H\lambda p_H\rho p_H + p_H\lambda p_H B_2 p_H(I-B_2'')^{-1}p_H\rho p_H.$$

It follows that $p_H x \in \mathfrak{S}$, and so $p_H R\Sigma^{-1} = \mathfrak{S}$, as wanted. □

Assume now that $x \in \ker(R\Sigma^{-1} \to \mathcal{R}) = \ker(R\Sigma^{-1} \to P^{\mathrm{rat}}_{\mathbf{K}}(E))$ and write x as in (2.4), with $B = B_1 + B_2$ as before. Then

$$x = \lambda(I-B_1)^{-1}\rho + \lambda(I-B_1)^{-1}B_2(I-B_2)^{-1}\rho. \tag{2.5}$$

Multiplying on the right by $1-p_H$, we get

$$x(1-p_H) = \lambda(I-B_1)^{-1}\rho(1-p_H) = \lambda(1-p_H)(I-B_1)^{-1}\rho(1-p_H) \in P_K(E)\Sigma(\epsilon_K)^{-1}$$

and $0 = \phi(x(1-p_H)) = \phi_K(x(1-p_H))$. Since ϕ_K is an isomorphism, we get $x(1-p_H) = 0$.

Hence we have

$$x = \lambda(I-B_1)^{-1}\rho_2 + \lambda(I-B_1)^{-1}B_2(I-B_2)^{-1}\rho_2, \tag{2.6}$$

where $\rho = \rho_1 + \rho_2$ with ρ_1 ending in $E^0 \setminus H$ and ρ_2 ending in H. By Claim 1 we have $p_H x = 0$, because ϕ is an isomorphism when restricted to $p_H R\Sigma^{-1}$. Now we are going to find a suitable expression for $x = (1-p_H)xp_H$. Write $\lambda = \lambda_1 + \lambda_2$ with $\lambda_1 = (1-p_H)\lambda$ and $\lambda_2 = p_H\lambda$. Then

$$(1-p_H)\lambda(I-B_1)^{-1}\rho_2 = \lambda_1(I-B_1)^{-1}\rho_2. \tag{2.7}$$

Similarly $(1-p_H)\lambda(I-B_1)^{-1}B_2(I-B_2)^{-1}\rho_2 = \lambda_1(I-B_1)^{-1}B_2(I-B_2)^{-1}\rho_2$. Write $B_2 = B_2' + B_2''$, with B_2' starting in $E^0 \setminus H$ and B_2'' starting in H. Then $B_2''B_2' = 0$ and $(I-B_2)^{-1} = (I-B_2')^{-1}(I-B_2'')^{-1}$, so that

$$\begin{aligned}(1-p_H)\lambda(I-B_1)^{-1}B_2(I-B_2)^{-1}\rho_2 &= \lambda_1(I-B_1)^{-1}B_2(I-B_2)^{-1}\rho_2\\ &= \lambda_1(I-B_1)^{-1}B_2(I+B_2')(I-B_2'')^{-1}\rho_2\\ &= \lambda_1(I-B_1)^{-1}B_2(I-B_2'')^{-1}\rho_2.\end{aligned} \tag{2.8}$$

From (2.7), (2.8) and (2.6) we get

$$x = (1-p_H)xp_H = \lambda_1(I-B_1)^{-1}\rho_2 + \lambda_1(I-B_1)^{-1}B_2(I-B_2'')^{-1}\rho_2. \tag{2.9}$$

It follows that $x \in \sum_{i=1}^k P^{\mathrm{rat}}_K(E/H)e_iP^{\mathrm{rat}}_{\mathbf{K}_H}(E_H)$, where $e_1,\dots,e_k$ is the family of *crossing edges*, that is, the family of edges $e \in E^1$ such that $s(e) \in E^0 \setminus H$ and $r(e) \in H$. Write $x = \sum_{i=1}^k\sum_{j=1}^{m_i} a_{ij}e_ib_{ij}$ for certain $a_{ij} \in P^{\mathrm{rat}}_K(E/H)$ and $b_{ij} \in P^{\mathrm{rat}}_{\mathbf{K}_H}(E_H)$. Then we have

$$0 = \phi(x) = \sum_{i=1}^k\sum_{j=1}^{m_i} a_{ij}e_ib_{ij},$$

this element being now in $P_{\mathbf{K}}((E))$. Clearly this implies that $\sum_{j=1}^{m_i} a_{ij}e_ib_{ij} = 0$ in $P_{\mathbf{K}}((E))$ for all $i = 1,\dots,k$. So the result follows from the following claim:

Claim 2. Let e be a crossing edge, so that $s(e) \in E^0\setminus H$ and $r(e) \in H$. Assume that $b_1,\dots,b_m \in r(e)P_{\mathbf{K}_H}((E_H))$ are K-linearly independent elements, and assume that $a_1e,\dots,a_me$ are not all 0, where $a_1,\dots,a_m \in P_K((E\setminus H))$. Then $\sum_{i=1}^m a_ieb_i \neq 0$ in $P_{\mathbf{K}}((E))$.

Proof of Claim 2. By way of contradiction, suppose that $\sum_{i=1}^m a_ieb_i = 0$. We may assume that $a_1e \neq 0$. Let γ be a path in the support of a_1 such that $r(\gamma) = s(e)$.

For every path μ with $s(\mu) = r(e)$ we have that the coefficient of $\gamma e\mu$ in $a_i e b_i$ is $a_i(\gamma)b_i(\mu)$, so that $\sum_{i=1}^m a_i(\gamma)b_i(\mu) = 0$ for every μ such that $s(\mu) = r(e)$. Since every path in the support of each b_i starts with $r(e)$, we get that

$$\sum_{i=1}^{m} a_i(\gamma)b_i = 0$$

with $a_1(\gamma) \neq 0$, which contradicts the linear independence over K of $b_1, \dots, b_m$. This concludes the proof of the theorem. □

Following [**6**, Section 2], we define, for $e \in E^1$, the right transduction $\tilde{\delta}_e \colon P_{\mathbf{K}}((E)) \to P_{\mathbf{K}}((E))$ corresponding to e by

$$\tilde{\delta}_e(\sum_{\alpha \in E^*} \lambda_\alpha \alpha) = \sum_{\substack{\alpha \in E^* \\ s(\alpha)=r(e)}} \lambda_{e\alpha}\alpha.$$

Note that the left transductions are not even defined on $P_{\mathbf{K}}((E))$ in general.

Observe that $R := P_{\mathbf{K}}(E)$ is closed under all the right transductions, i.e. $\tilde{\delta}_e(R) \subseteq R$. Some of the proofs in [**6**] make use of the fact that the usual path algebra $P_K(E)$ is closed under *left and right* transductions. Fortunately we have been able to overcome the potential problems arising from the failure of invariance of R under left transductions by using alternative arguments.

We are now ready to get a description of the algebra $Q_{\mathbf{K}}(E)$ as a universal localization of the path algebra $P_{\mathbf{K}}(E)$.

Write $R := P_{\mathbf{K}}(E)$. For any $v \in E^0$ such that $s^{-1}(v) \neq \emptyset$ we put $s^{-1}(v) = \{e_1^v, \dots, e_{n_v}^v\}$, and we consider the left R-module homomorphism

$$\begin{aligned} \mu_v \colon Rv &\longrightarrow \bigoplus_{i=1}^{n_v} Rr(e_i^v) \\ r &\longmapsto \left(re_1^v, \dots, re_{n_v}^v\right) \end{aligned}$$

Write $\Sigma_1 = \{\mu_v \mid v \in E^0,\ s^{-1}(v) \neq \emptyset\}$.

THEOREM 2.3. *Let* $\mathbf{K}$ *and* E *be as before. Let* Σ *denote the set of matrices over* $P_{\mathbf{K}}(E)$ *that are sent to invertible matrices by* ϵ *and let* Σ_1 *be the set of maps defined above. Then we have that* $Q_{\mathbf{K}}(E) := P_{\mathbf{K}}(E)(\Sigma \cup \Sigma_1)^{-1}$ *is a hereditary von Neumann regular ring and all finitely generated projective* $Q_{\mathbf{K}}(E)$*-modules are induced from* $P_{\mathbf{K}}^{\mathrm{rat}}(E)$*. Moreover each element of* $Q_{\mathbf{K}}(E)$ *can be written as a finite sum*

$$\sum_{\gamma \in \mathrm{Path}(E)} a_\gamma \gamma^*,$$

where $a_\gamma \in P_{\mathbf{K}}^{\mathrm{rat}}(E)r(\gamma)$.

PROOF. First observe that $P_{\mathbf{K}}(E)$ is a hereditary ring and that $\mathcal{V}(P_{\mathbf{K}}(E)) = (\mathbb{Z}^+)^d$, where $|E^0| = d$. This follows by successive use of [**11**, Theorem 5.3].

In order to get that the right transduction $\tilde{\delta}_e \colon P_{\mathbf{K}}((E)) \to P_{\mathbf{K}}((E))$ corresponding to e is a right τ_e-derivation, that is,

$$\tilde{\delta}_e(rs) = \tilde{\delta}_e(r)s + \tau_e(r)\tilde{\delta}_e(s) \tag{2.10}$$

for all $r, s \in P_{\mathbf{K}}((E))$, we have to modify slightly the definition of τ_e given in [**6**, page 220]. Concretely we define τ_e as the endomorphism of $P_{\mathbf{K}}((E))$ given by the composition

$$P_{\mathbf{K}}((E)) \to \prod_{v\in E^0} K_{[v]}v \to \prod_{v\in E^0} K_{[v]}v \to P_{\mathbf{K}}((E)),$$

where the first map is the augmentation homomorphism, the third map is the canonical inclusion, and the middle map is the K_0-lineal map given by sending $s(e)$ to $r(e)$, and any other idempotent v with $v \neq s(e)$ to 0. Observe that this restricts to an endomorphism of $P_{\mathbf{K}}(E)$ and that the proof in [**6**, Lemma 2.4] gives the desired formula (2.10) for $r, s \in P_{\mathbf{K}}((E))$ and, in particular for $r, s \in P_{\mathbf{K}}(E)$.

Note that, by the argument given after the proof of Proposition 2.7 in [**6**], the algebras $P_{\mathbf{K}}^{\mathrm{rat}}(E)$ are stable under all the right transductions. Hence, the constructions in [**6**, Section 2] apply to $R := P_{\mathbf{K}}^{\mathrm{rat}}(E)$ (with some minor changes), and we get that $R\Sigma_1^{-1} = R\langle\overline{E};\tau,\tilde{\delta}\rangle/I$, where I is the ideal of $R\langle\overline{E};\tau,\tilde{\delta}\rangle$ generated by the idempotents $q_v := v - \sum_{e\in s^{-1}(v)} ee^*$ for $v \notin \mathrm{Sink}(E)$.

By Theorem 2.2, we have

$$\begin{aligned} Q_{\mathbf{K}}(E) &= P_{\mathbf{K}}(E)(\Sigma\cup\Sigma_1)^{-1} = (P_{\mathbf{K}}(E)\Sigma^{-1})\Sigma_1^{-1} \\ &= P_{\mathbf{K}}^{\mathrm{rat}}(E)\Sigma_1^{-1} = (P_{\mathbf{K}}^{\mathrm{rat}}(E))\langle\overline{E};\tau,\tilde{\delta}\rangle/I. \end{aligned}$$

By a result of Bergman and Dicks [**12**] any universal localization of a hereditary ring is hereditary, thus we get that both $P_{\mathbf{K}}^{\mathrm{rat}}(E)$ and $Q_{\mathbf{K}}(E)$ are hereditary rings. Since $P_{\mathbf{K}}^{\mathrm{rat}}(E)$ is hereditary, closed under inversion in $P_{\mathbf{K}}((E))$ (by Theorem 2.2), and closed under all the right transductions $\tilde{\delta}_e$, for $e \in E^1$, the proof of [**6**, Theorem 2.16] gives that $Q_{\mathbf{K}}(E)$ is von Neumann regular and that every finitely generated projective module is induced from $P_{\mathbf{K}}^{\mathrm{rat}}(E)$.

The last statement follows from [**6**, Proposition 2.5(ii)]. This concludes the proof of the theorem. □

REMARK 2.1. Theorem 2.16 in [**6**] is stated for a subalgebra R of $P_K((E))$ which is closed under all left and right transductions (and which is inversion closed in $P_K((E))$). However the invariance under right transductions is only used in the proof of that result to ensure that the ring R is left semihereditary. Since we are using the opposite notation concerning (CK1) and (CK2), the above hypothesis translates in our setting into the condition that $P_{\mathbf{K}}(E)$ and $P_{\mathbf{K}}^{\mathrm{rat}}(E)$ should be invariant under all *left* transductions, which is not true in general as we observed above. We overcome this problem by the use of the result of Bergman and Dicks ([**12**]), which guarantees that $P_{\mathbf{K}}(E)$ and $P_{\mathbf{K}}^{\mathrm{rat}}(E)$ are indeed right and left hereditary (see the proof of Theorem 2.3).

For our last result, we need an auxiliary lemma.

LEMMA 2.4. Let $(I,\leq)$ be a finite poset which is a tree, with maximum element i_0. Let $\mathbf{K} = \{K_i\}_{i\in I}$ be a poset of fields, so that $K_i \subseteq K_j$ if $j \leq i$ in I. Then there exists a field K and an embedding of $\mathbf{K}$ into the constant system K over I, i.e., there is a collection of field morphisms $\varphi_i\colon K_i \to K$ such that $\varphi_i|_{K_j} = \varphi_j$ for all $i \leq j$.

PROOF. For a finite poset I as in the statement, define the *depth* of I as the maximum of the lengths of maximal chains $i_k < i_{k-1} < \cdots < i_0$. We prove the result by induction on the depth of I. If the depth of I is 0, there is nothing to prove. Assume the result is true for finite posets of depth at most r satisfying the hypothesis in the statement and let $(I, \leq)$ such a finite poset of depth $r+1$, where $r \geq 0$. Write $\mathrm{L}(I, i_0) = \{a_1, \dots, a_n\}$. Then the depth of each of the posets $I \downarrow a_t$ is at most r, so that there are embeddings $\psi_t \colon \mathbf{K}_{I\downarrow a_t} \to L_t$ for some fields L_t, for $t = 1, \dots, r$. This means that there are field embeddings $\psi_{t,i} \colon K_i \to L_t$ for each $i \leq a_t$, such that $\psi_{t,i}|_{K_j} = \psi_{t,j}$ for $i \leq j \leq a_t$. Now, setting $K_0 := K_{i_0}$, there exist a field K and embeddings $\delta_t \colon L_t \to K$ for $t = 1, \dots, n$ such that $\delta_t \circ ((\psi_{t,a_t})|_{K_0}) = \delta_{t'} \circ (\psi_{t',a_{t'}})|_{K_0})$ for $1 \leq t, t' \leq n$. Define $\varphi_i \colon K_i \to K$ to be $\delta_t \circ \psi_{t,i}$ if $i \leq a_t$. (Observe that this is well-defined by our hypothesis on I.) Finally define $\varphi_{i_0} \colon K_0 \to K$ as $\varphi_{i_0} := \delta_t \circ ((\psi_{t,a_t})|_{K_0})$, which is independent of the choice of t by the above.

If $i \leq j \leq a_t$ for some t, then

$$\varphi_i|_{K_j} = (\delta_t \circ \psi_{t,i})|_{K_j} = \delta_t \circ (\psi_{t,i}|_{K_j}) = \delta_t \circ \psi_{t,j} = \varphi_j.$$

If $i < i_0$, then there is a unique $t \in \{1, \dots, n\}$ such that $i \leq a_t < i_0$, and so

$$\varphi_i|_{K_0} = (\delta_t \circ \psi_{t,i})|_{K_0} = \delta_t \circ (\psi_{t,i}|_{K_{a_t}})|_{K_0} = \delta_t \circ (\psi_{t,a_t})|_{K_0} = \varphi_{i_0}.$$

This completes the proof. □

Define the Leavitt path algebra $L_{\mathbf{K}}(E)$ associated to the poset of fields $\mathbf{K}$ and the finite quiver E as the universal localization of $P_{\mathbf{K}}(E)$ with respect to the set Σ_1. Let $M(E)$ be the abelian monoid with generators E^0 and relations given by $v = \sum_{e \in s^{-1}(v)} r(e)$, see [**10**] and [**1**, Chapter 3].

THEOREM 2.5. With the above notation, we have natural isomorphisms

$$M(E) \cong \mathcal{V}(L_{\mathbf{K}}(E)) \cong \mathcal{V}(Q_{\mathbf{K}}(E)).$$

PROOF. The proof that $M(E) \cong \mathcal{V}(L_{\mathbf{K}}(E))$ follows from Bergman's results [**11**], as in [**10**, Theorem 3.5].

Note that $R := P^{\mathrm{rat}}_{\mathbf{K}}(E)$ is semiperfect. Thus we get $\mathcal{V}(R) \cong (\mathbb{Z}^+)^{|E_0|}$ in the natural way, that is the generators of $\mathcal{V}(R)$ correspond to the projective modules Rv for $v \in E^0$. By Theorem 2.3, we get that the natural map $M(E) \to \mathcal{V}(Q_{\mathbf{K}}(E))$ is surjective. To show injectivity, let K and $\{\varphi_i\}_{i \in I}$ be a field and maps satisfying the hypothesis of Lemma 2.4. Then it is easily seen that there is a unital ring homomorphism $\varphi \colon Q_{\mathbf{K}}(E) \to Q_K(E)$. Indeed, one directly verifies that there is well-defined homomorphism $\varphi \colon P_{\mathbf{K}}(E) \to P_K(E)$ given by the rule

$$\varphi(\sum a_\gamma \gamma) = \sum \varphi_{[r(\gamma)]}(a_\gamma)\gamma.$$

Let Σ (respectively Σ_K) be the set of all square matrices over $P_{\mathbf{K}}(E)$ (respectively over $P_K(E)$) which are sent to invertible matrices through the augmentation map ϵ. Since $\epsilon(\varphi(A)) = \varphi(\epsilon(A))$, it is obvious that $\varphi(\Sigma) \subseteq \Sigma_K$, and consequently the map φ can be uniquely extended to a homomorphism (also denoted by φ) from $Q_{\mathbf{K}}(E) = P_{\mathbf{K}}(E)(\Sigma \cup \Sigma_1)^{-1}$ to $Q_K(E) = P_K(E)(\Sigma_K \cup \Sigma_1)^{-1}$. By [**6**, Theorem 4.2], we have $M(E) \cong \mathcal{V}(Q_{K_0}(E))$ and $M(E) \cong \mathcal{V}(Q_K(E))$ canonically, so that we get

$$M(E) \cong \mathcal{V}(Q_{K_0}(E)) \longrightarrow \mathcal{V}(Q_{\mathbf{K}}(E)) \longrightarrow \mathcal{V}(Q_K(E)) \cong M(E),$$

and the composition of the maps above is the identity. It follows that the map $M(E) \to \mathcal{V}(Q_{\mathbf{K}}(E))$ is injective and so it must be a monoid isomorphism. □

We adopt now the specific setting of [**4**] for our graphs in order to obtain a property of $Q_{\mathbf{K}}(E)$ needed in [**4**]. So in addition to our previous assumptions on E, we make the following requirements:

(1) There is a partition $I = I_{\text{free}} \bigsqcup I_{\text{reg}}$ (where $I = E/{\sim}$).
(2) For each $v \in E^0$ such that $[v]$ is not minimal in I and $[v] \in I_{\text{free}}$, we have that $E[v]$ is a graph with a single vertex v and a single arrow α^v, with $s(\alpha(v)) = r(\alpha(v) = v$.
(3) For each $v \in E^0$ such that $[v] \in I_{\text{reg}}$, we have that $E[v]$ is a graph with at least two edges.
(4) If $[v]$ is minimal in I, then either v is a sink or $[v] \in I_{\text{reg}}$.

With these conditions at hand, we finally observe that we can obtain $Q_{\mathbf{K}}(E)$ by inverting a smaller set of matrices over $P_{\mathbf{K}}(E)$. For each non-minimal $[v] \in I_{\text{free}}$, let

$$\Sigma_i := \{p(\alpha^v) \mid p(x) \in K_{[v]}[x] \text{ and } p(0) \neq 0\}.$$

be the set of polynomials in α^v with coefficients in $K_{[v]}$ with nonzero constant term. For each $i \in I_{\text{reg}}$, set $p_i = \sum_{v \in i} v$ and let Σ_i be the set of square matrices M over $p_i P_{K_i}(E) p_i$ such that $\epsilon_i(M)$ is an invertible matrix over $\prod_{v\in i} K_i$, where $p_i P_{K_i}(E) p_i \to \prod_{v\in i} K_i$ is the augmentation homomorphism. Finally set $\Sigma' = \bigsqcup_{i\in I} \Sigma'_i$, where, for $i \in I$, $\Sigma'_i = \{M + (1 - p_i)I_m \mid M \in M_m(P_{\mathbf{K}}(E)) \cap \Sigma_i\}$.

PROPOSITION 2.6. Let E and Σ, Σ' be as above. Then we have $Q_{\mathbf{K}}(E) = L_{\mathbf{K}}(E)(\Sigma')^{-1}$.

PROOF. Since $Q_{\mathbf{K}}(E) = L_{\mathbf{K}}(E)\Sigma^{-1}$, we only have to show that every element of Σ is invertible over $L_{\mathbf{K}}(E)(\Sigma')^{-1}$. For this, it is enough to show that any matrix of the form $I - A$ is invertible over $L_{\mathbf{K}}(E)(\Sigma')^{-1}$, where A is a square matrix over $P_{\mathbf{K}}(E)$ such that $\epsilon(A) = 0$.

Now given such a matrix A, we can uniquely write it in the form

$$A = A_0 + B + A_1 + \cdots + A_l,$$

where $\mathrm{L}(I, i_0) = \{i_1, \ldots, i_l\}$, $\epsilon(B) = \epsilon(A_k) = 0$ for $k = 0, \ldots, l$, A_0 is a square matrix over $p_{i_0} P_{\mathbf{K}}(E) p_{i_0}$, all paths in the support of B start in a vertex in the component of i_0 and end in a vertex in the component of some $i \in I$, where $i < i_0$, and each A_k is a square matrix over $P_{\mathbf{K}_{i_k}}(E_{i_k})$. Here, E_{i_k} is the hereditary subset $E_{I\downarrow i_k}$ and $\mathbf{K}_{i_k}$ is the restriction of the poset of fields $\mathbf{K}$ to $I \downarrow i_k$.

Observe that $(B + \sum_{k=1}^{l} A_k)A_0 = 0$, so that we obtain

$$(I - A)^{-1} = (I - A_0)^{-1}(I - (B + \sum_{k=1}^{l} A_k))^{-1}$$

in any ring where the matrices on the right hand side are invertible. Now observe that $(\sum_{k=1}^{l} A_k)B = 0$, because all the paths in the support of B start at the component i_0, so we get

$$(I - (B + \sum_{k=1}^{l} A_k))^{-1} = (I - B)^{-1}(I - \sum_{k=1}^{l} A_k)^{-1}$$

in any ring where the matrices on the right hand side are invertible. Now since

$$(I \downarrow i_k) \cap (I \downarrow i_{k'}) = \emptyset$$

for $k \neq k'$ we get that

$$(I - \sum_{k=1}^{l} A_k)^{-1} = \prod_{k=1}^{l} (I - A_k)^{-1}$$

again in any ring where all matrices on the right hand side are invertible.

Continuing in this way, we obtain that the matrix $(I-A)^{-1}$ can be expressed as a product of matrices of the forms $(I-A')^{-1}$, where A' is a matrix over $p_i P_{K_i}(E) p_i$ such that $\epsilon(A')=0$, and $(I-B')^{-1}$, where B' is a matrix such that all paths in the support of B' start at a vertex in a fixed component $i \in I$ and end at a component which is strictly less than i. Now observe that $(B')^2=0$ and so $(I-B')^{-1}=I+B'$ is a matrix over $P_{\mathbf{K}}(E)$. Therefore, since all the matrices $I-A'$ and $I-B'$ as above are invertible in $L_{\mathbf{K}}(E)(\Sigma')^{-1}$, we see that $I-A$ is also invertible, and the proof is complete. (Note that, in case $i=[v] \in I_{\text{free}}$ is non-minimal, the determinant of the matrix $Iv-A'$ is of the form $v+p(\alpha_v)$, where $p(0)=0$, so the matrix $I-A'$ is invertible over $L_{\mathbf{K}}(E)(\Sigma')^{-1}$.) □

References

[1] G. Abrams, P. Ara, and M. Siles Molina, *Leavitt path algebras*, Lecture Notes in Mathematics, vol. 2191, Springer, London, 2017. MR3729290

[2] P. Ara, *The realization problem for von Neumann regular rings*, Ring theory 2007, World Sci. Publ., Hackensack, NJ, 2009, pp. 21–37, DOI 10.1142/9789812818331_0004. MR2513205

[3] P. Ara, *The regular algebra of a poset*, Trans. Amer. Math. Soc. **362** (2010), no. 3, 1505–1546, DOI 10.1090/S0002-9947-09-04884-3. MR2563739

[4] P. Ara, J. Bosa, E, Pardo, The realization problem for finitely generated refinement monoids, arXiv:1907.03648 [math.RA], 2019.

[5] P. Ara, J. Bosa, E. Pardo, A. Sims, The groupoids of adaptable separated graphs and their type semigroup, arXiv:1904.05197 [math.OA], 2019.

[6] P. Ara and M. Brustenga, *The regular algebra of a quiver*, J. Algebra **309** (2007), no. 1, 207–235, DOI 10.1016/j.jalgebra.2006.10.013. MR2301238

[7] P. Ara, M. Brustenga, Mixed quiver algebras, arXiv:0909.0421 [math.RA], 2009.

[8] P. Ara and K. R. Goodearl, *The realization problem for some wild monoids and the Atiyah problem*, Trans. Amer. Math. Soc. **369** (2017), no. 8, 5665–5710, DOI 10.1090/tran/6889. MR3646775

[9] P. Ara, K. R. Goodearl, K. C. O'Meara, and E. Pardo, *Separative cancellation for projective modules over exchange rings*, Israel J. Math. **105** (1998), 105–137, DOI 10.1007/BF02780325. MR1639739

[10] P. Ara, M. A. Moreno, and E. Pardo, *Nonstable K-theory for graph algebras*, Algebr. Represent. Theory **10** (2007), no. 2, 157–178, DOI 10.1007/s10468-006-9044-z. MR2310414

[11] G. M. Bergman, *Coproducts and some universal ring constructions*, Trans. Amer. Math. Soc. **200** (1974), 33–88, DOI 10.2307/1997247. MR0357503

[12] G. M. Bergman and W. Dicks, *Universal derivations and universal ring constructions*, Pacific J. Math. **79** (1978), no. 2, 293–337. MR531320

[13] P. M. Cohn, *Free ideal rings and localization in general rings*, New Mathematical Monographs, vol. 3, Cambridge University Press, Cambridge, 2006. MR2246388

[14] K. R. Goodearl, *von Neumann regular rings*, 2nd ed., Robert E. Krieger Publishing Co., Inc., Malabar, FL, 1991. MR1150975

[15] K. R. Goodearl, *von Neumann regular rings and direct sum decomposition problems*, Abelian groups and modules (Padova, 1994), Math. Appl., vol. 343, Kluwer Acad. Publ., Dordrecht, 1995, pp. 249–255. MR1378203

[16] W. Lück, L^2-invariants: theory and applications to geometry and K-theory. Ergebnisse der Mathematik und ihrer Grenzgebiete, 44. Springer-Verlag, Berlin, 2002.

[17] S. W. Rigby, The groupoid approach to Leavitt path algebras. Preprint, 2018.

[18] A. H. Schofield, *Representation of rings over skew fields*, London Mathematical Society Lecture Note Series, vol. 92, Cambridge University Press, Cambridge, 1985. MR800853

Departament de Matemàtiques, Universitat Autònoma de Barcelona, 08193, Bellaterra (Barcelona), Spain

Email address: `para@mat.uab.es, mbrusten@mat.uab.es`

Contemporary Mathematics
Volume **738**, 2019
https://doi.org/10.1090/conm/738/14875

Roots of elements in Lie groups and the exponential maps

S. G. Dani

ABSTRACT. Given a group G and a natural number k not all elements of G may admit a k-th root in G. When G is a connected Lie group, an element admits roots of all orders if and only if it is contained in a (continuous) one-parameter subgroup of G, namely in the image of the exponential map associated with G, and when this holds the element is said to be exponential. Here we discuss various issues concerning the sets of exponential elements in connected Lie groups, and in particular Lie groups in which the set is the whole, they being called exponential groups.

1. Introduction

Let G be a group and $g \in G$. For $k \in \mathbb{N}$, $k \geq 2$, an element $x \in G$ is called a kth root of g if $x^k = g$. We note that in a general group most elements may not admit roots of a given order. For example, if $G = \mathbb{Z}^n$ for some $n \geq 1$, where $\mathbb{Z}$ denotes the group of integers, then $g \in G$ has kth roots only for finitely many ks. On the other hand, if $G = \mathbb{R}^n$ or $\mathbb{T}^n$, $n \geq 1$, where $\mathbb{R}$ stands for the group of real numbers and $\mathbb{T}$ for the circle group $\mathbb{R}/\mathbb{Z}$, then every element of G has a k-th root for all $k \in \mathbb{N}$. The latter are Lie groups and in fact any connected abelian Lie group is a Cartesian products groups of this form, and hence in such a group G every $g \in G$ admits a k-th root for all k.

In a more general Lie group, while there is a rich class of elements admitting roots of all orders, there can also be large classes of elements for which this does not hold. Moreover, the issue of understanding when it does hold remains open, though there is a vast amount of literature on the topic, thanks to contributions of many authors. Our aim here is give the reader a flavour of the topic, and also to present, in the second part of the paper, a modified simpler proof of an earlier unpublished result of the present author (see Theorem 3.1 for details).

For wider accessibility of the contents of the paper we begin with a relatively non-technical overview of Lie groups.

2. A brief introduction to Lie groups

Consider the group $\mathrm{GL}(n, \mathbb{R})$ consisting of all $n \times n$ nonsingular matrices. It is an open subset of the space of $n \times n$ matrices with real entries, which is itself a Euclidean space (of dimension n^2). In particular it is a *manifold*, namely, a locally

---

2010 *Mathematics Subject Classification.* 22E15.

euclidean space. It is also a group under matrix multiplication, and the group operations of multiplication and inversion are continuous. Such a group, which is a manifold and the group operations are continuous, is called a Lie group. We note that $\mathrm{GL}(n,\mathbb{R})$ is not connected. On the other hand the subgroup $\mathrm{GL}(n,\mathbb{R})^+$ consisting of all its elements with positive determinant is connected. It is an open submanifold of $\mathrm{GL}(n,\mathbb{R})$ and is thus a connected Lie group. Here we shall be concerned with *connected* Lie groups.

2.1. Linear Lie groups. Closed subgroups of $\mathrm{GL}(n,\mathbb{R})$, and more generally of any Lie group, also turn out to be Lie groups. As a general statement this is a deep result, due to Cartan, but can be confirmed easily in many examples such as the special linear group $\mathrm{SL}(n,\mathbb{R})$, the orthogonal groups, symplectic groups, groups of upper triangular matrices with positive diagonal entries, upper triangular matrices with 1s on the diagonal, etc.. A closed subgroup of $\mathrm{GL}(n,\mathbb{R})$, $n\in\mathbb{N}$, will be called a *linear* group.

The group $\mathrm{GL}(n,\mathbb{C})$ consisting of $n\times n$ nonsingular matrices with entries in $\mathbb{C}$, the field of complex numbers, is also evidently a Lie group; we note that it can be also realized canonically as a closed subgroup of $\mathrm{GL}(2n,\mathbb{R})$. Closed subgroups of $\mathrm{GL}(n,\mathbb{C})$, in particular those analogous to the subgroups of $\mathrm{GL}(n,\mathbb{R})$ noted above, are also Lie groups.

It is an interesting fact, a theorem of Ado, that for any connected Lie group G there exists a linear group H such that G and H can be realized as quotients of a common Lie group L, by countable subgroups of L that are contained in its center, namely L is a covering group for both G and H. In view of this for many purposes it suffices to study the issue under consideration for linear Lie groups, at least in the initial instance.

2.2. The exponential map. Let $M(n,\mathbb{R})$ denote the vector space of $n\times n$ matrices with entries in $\mathbb{R}$. Then we have the *exponential map* $\mathrm{Exp}: M(n,\mathbb{R})\to \mathrm{GL}(n,\mathbb{R})$ defined by

$$\mathrm{Exp}(X)=I+\sum_{j=1}^{\infty}\frac{1}{j!}X^j.$$

Let $G\subset \mathrm{GL}(n,\mathbb{R})$ be a linear connected Lie group and let

$$\mathfrak{G}=\{X\in M(n,\mathbb{R})\mid \mathrm{Exp}(tX)\in G \text{ for all } t\in\mathbb{R}\}.$$

Then $\mathfrak{G}$ is a vector subspace of $M(n,\mathbb{R})$; in fact it is also a Lie subalgebra of $M(n,\mathbb{R})$, viz. for all $X,Y\in\mathfrak{G}$, the element $XY-YX$, which is called the Lie product of X and Y and is denoted by $[X,Y]$, is also contained in $\mathfrak{G}$. We thus get a natural map $\mathrm{Exp}:\mathfrak{G}\to G$; it is called the *exponential map* of G.

Analogously, with each Lie group G is associated a Lie algebra $\mathfrak{G}$, in fact a Lie subalgebra of $M(n,\mathbb{R})$ for *some* n, and an exponential map of $\mathfrak{G}$ into G. Moreover, if G and H are connected Lie groups with the associated Lie algebras $\mathfrak{G}$ and $\mathfrak{H}$ respectively, and $\theta: G\to H$ is a continuous homomorphism, then there is a unique linear map (in fact a Lie algebra homomorphism) $\theta':\mathfrak{G}\to\mathfrak{H}$ such that $\exp(\theta'(X))=\theta(\exp(X))$ for all $X\in\mathfrak{G}$. When G is a covering group of H the Lie algebras of G and H can be identified canonically in such a way that θ' is the identity map.

Now consider a connected Lie group G. An element $g\in G$ is said to be *exponential* if it is contained in the image of the exponential map of G. We note that if $g = \exp(X)$ for some $X \in \mathfrak{G}$, the Lie algebra of G, then $(\exp(X/k))^k = \exp(X) = g$; this relation follows from a straightforward computation using the definition in the case of linear groups - for a general Lie group it is analogously inbuilt in the definition of the exponential map (which we did not go into here), but can also be seen from the interrelations of a general (connected) Lie group with a linear Lie group that we noted above. This shows that (in any Lie group) any exponential element admits roots of all orders. Interestingly the converse also holds, as was proved by M. McCrudden.

THEOREM 2.1. *Let G be a connected Lie group and $g \in G$. Then g admits a kth root for all $k \in \mathbb{N}$ if and only if g is exponential.*

2.3. Exponential and weakly exponential groups. A Lie group is said to be *exponential* if all its elements are exponential, namely if its exponential map Exp is surjective. It is said to be *weakly exponential* if the image of Exp is dense in G. In view of the Theorem 2.1 these properties relate nicely to understanding roots in Lie groups. We recall the following in this respect, for connected Lie groups G.

- If G is abelian it has the form $\mathbb{R}^n \times \mathbb{T}^m$, $m, n \geq 0$, and hence it is exponential.
- If G is a nilpotent Lie group then it is exponential, as can be deduced from the abelian case, using induction.
- The group $\mathrm{GL}(n, \mathbb{C})$ is exponential, as can be confirmed using Jordan canonical form of matrices.
- Any compact (connected) Lie group, e.g. the special orthogonal group, is exponential.
- The groups $\mathrm{GL}(n, \mathbb{R})^+$ and $\mathrm{SL}(n, \mathbb{R})$ are **not** exponential: a diagonal matrix $\mathrm{diag}(a_1, \ldots, a_n)$, with $a_j \in \mathbb{R}^*(= \mathbb{R}\backslash\{0\})$ pairwise distinct, has a square root only if all a_j are positive. Since these elements and their conjugates in the group form an open dense subset it follows also that these groups are not even weakly exponential.
- $\mathrm{PSL}(2, \mathbb{R})(= \mathrm{SL}(2, \mathbb{R})/\{\pm I\}$, where I is the identity matrix) is exponential. However, $\mathrm{PSL}(n, \mathbb{R})$ are not exponential for $n \geq 3$, and also not weakly exponential. On the other hand the groups $\mathrm{PSL}(n, \mathbb{C})$ are exponential for all $n \geq 2$.
- A semidirect product $G = A \cdot N$, where A is an abelian Lie group and N is a connected nilpotent Lie group, via an action of A on N by automorphisms, is exponential if and only if for any $n \in N$ the subgroup of A consisting of elements fixing n is connected. However, all these groups are weakly exponential.

2.4. Structure of Lie groups. In order to give a flavour of the general case we recall some facts about the structure of a general Lie group. Let G be a connected Lie group. Then G has a unique maximal solvable connected normal Lie subgroup, called the *radical* of G (sometimes also called the *solvable radical*); it is a closed subgroup. For example a semidirect product $G = S \cdot R$, where $S = \mathrm{SL}(n, \mathbb{R})$ and R is a solvable connected Lie group, then R is the radical of G (as $\mathrm{SL}(n, \mathbb{R})$ has no nontrivial connected normal solvable Lie subgroup; here R may in particular be taken to be $\mathbb{R}^n$, and the action of S on R given as the natural action by linear transformations, in which case G is a group of affine automorphisms of $\mathbb{R}^n$, with the radical consisting of translations.

A Lie group G is said to be *semisimple* if the radical of G is trivial. The groups $\mathrm{SL}(n,\mathbb{R})$, $\mathrm{SL}(n,\mathbb{C})$, the symplectic group, the special orthogonal groups, are examples of semisimple Lie groups, and so are their covering groups and quotients by subgroups contained in the center; in particular quotients by the center, denoted by $\mathrm{PSL}(n,\mathbb{R})$ and $\mathrm{PSL}(n,\mathbb{C})$ in the case of $\mathrm{SL}(n,\mathbb{R})$ and $\mathrm{SL}(n,\mathbb{C})$ respectively, are semisimple Lie groups, and it may also be noted that their centers are trivial.

It is easy to see that if G is a connected Lie group and R is the radical of G then the quotient G/R is a semisimple Lie group. On the other hand it is also known that any connected Lie group G can be expressed as $G = S \cdot R$, where R is the radical, S is a connected semisimple Lie subgroup of G, and $S \cap R$ is a discrete subgroup contained in the center of G; thus $G = S \cdot R$ is an "almost" semidirect product, and it is indeed a semidirect product when the center of S is trivial. The decomposition of G as $S \cdot R$ as above is called the *Levi decomposition* of G. We note that the groups S and R together with the action of S on R determine the group G.

2.5. General results on exponential and weakly exponential groups. In view of the Levi decomposition, one strategy for understanding issues such as exponentiality or weak exponentiality for a general Lie group G would be to understand them separately for semisimple and solvable groups, and also analyse the general case in terms of the action of S on R, with S and R as in the Levi decomposition of G. It is also possible, of course, to analyse the question without recourse to the Levi decomposition, though other means, and the reader may benefit referring [**2**], Corollary D, in this respect. From our present perspective however, relating the issues to the Levi decomposition, thus connecting them to the broad features of the Lie group would be pertinent.

There is a complete classification of semisimple Lie algebras, namely those with no solvable Lie ideals. To each semisimple Lie algebra there corresponds a unique simply connected Lie group with that as the corresponding Lie algebra, and every connected semisimple Lie group is a quotient of one of these, by a subgroup contained in its (discrete) centre. This provides a complete description of the class of all connected semisimple Lie groups. The questions of exponentiality and weak exponentiality have been analysed for this class and groups for which these properties hold have been catalogued. The work is highly technical and we shall not go into it here. (The interested reader is referred to [**7**], [**11**], and other references cited there.)

For solvable Lie groups one does not expect a cataloguing, as there are quite flexible ways of constructing such groups. The examples introduced in the last bullet item in § 2.3 in some ways form a prototype, but the class includes in particular their Lie subgroups, which need not be of the same type. As with the examples in § 2.3, all solvable (connected) Lie groups are known to be weakly exponential. Also, good necessary and sufficient conditions are known for exponentiality of these groups, by results of Wüstner (see [**7**], Theorem 4.2). Some of the main criteria are in terms of the actions of Cartan subgroups of the given group on the nilradical (the maximal connected nilpotent normal Lie subgroup), which in the case of the examples as above correspond to the A action on N.

The second part of the strategy, for a general connected Lie group G, is to relate the questions involved to those of the subgroups S and R as in the Levi decomposition; recall that $G = S \cdot R$, with S semisimple, R the radical. It is known that G is weakly exponential if and only if S is weakly exponential (see [**7**], Theorem 3.1). With regard to the issue of exponentiality of such a mixed G, however, there is not enough clarity; when G is a semidirect product of S and R then G being exponential of course implies S to be exponential. On the other hand G being exponential does not imply that the radical R is exponential. An example of such a exponential group G, with $S = \mathrm{SU}(2, \mathbb{C})$ and R is a group which is a semidirect product of $\mathbb{R}$ with $\mathbb{C}^2$ which is not exponential, is described in [**10**]. On the other hand, it was proved in [**4**] that under a certain condition on S, G being exponential implies that R is exponential:

THEOREM 2.2. *Let G be an exponential Lie group and R be the radical of G. Suppose that there exists a unipotent one-parameter subgroup U of G/R such that the centralizer of U in G/R does not contain a compact subgroup of positive dimension. Then R is an exponential Lie group.*

The condition on the semisimple group G/R in the above hypothesis holds for all complex semisimple Lie groups, and more generally for what are known as *quasi-split* groups; we may recall here, without going into details, that if $G = KAN$ is an Iwasawa decomposition of G, where K is an Ad-compact subgroup, A is a maximal subgroup whose action on the Lie algebra is diagonalizable over $\mathbb{R}$ and N is a simply connected nilpotent Lie group, G is quasi-split if the centralizer of A in K is abelian.

The author was led to considering the question as in Theorem 2.2 in trying to understand the following result of Moskowitz and Sacksteder (in the case of complex algebraic groups it was proved earlier in [**2**], Corollary 6.4) which can now be deduced as a consequence of the theorem, using the fact that the center of an exponential solvable Lie group is connected.

COROLLARY 2.3. *Let G be a complex Lie group. If G is exponential then the center of G is connected.*

3. Criteria for exponentiality of elements and cosets

In the last section we discussed conditions for *Lie groups* to be exponential, namely all elements of the group to be exponential, which has been the focus of much of the study on the topic. In general it would be convenient to have general criteria for exponentiality of individual elements, or all elements in certain special subsets, like cosets of a subgroup. Apart from their independent interest the questions also have applications in determining interrelations between exponential groups; in particular our proof of Theorem 2.2 may be mentioned as one such application.

In this section we discuss certain methods and results in this respect.

3.1. Using Jordan decomposition. Let G be a linear connected Lie group. We recall that an element of g is said to be semisimple if it is diagonalizable as a matrix over the field of complex numbers, and is said to be unipotent if all its eigenvalues are 1. We say that G admits Jordan decomposition if for every $g \in G$ there exist a semisimple element s and a unipotent element u such that $g = su = us$; the decomposition, when it exists, is unique, as no nontrivial element can be both semisimple and unipotent. Most Lie groups of interest, in particular the algebraic subgroups of $\mathrm{GL}(n, \mathbb{R})$ (subgroups for which there exist polynomials in the

n^2 variables corresponding to the entries, such that the subgroup consists precisely of those elements at which the polynomials vanish), admit Jordan decomposition; this corresponds to the Jordan decomposition in $\mathrm{GL}(n, \mathbb{R})$ in the usual sense of Linear algebra. Moreover, in these groups every unipotent element is contained in a unique one-parameter subgroup, and in particular is an exponential element.

Now let G be a linear Lie group which admits Jordan decomposition, and every unipotent element is exponential. Let $g \in G$ and, and $g = su$ be its Jordan decomposition, where s is semisimple and u is unipotent. Suppose that g is contained in a one-parameter subgroup $\{g_t\}$, say $g_1 = g$. Let $g_t = s_t u_t$ be the Jordan decomposition of g_t, $t \in \mathbb{R}$. Then $s_1 = s$, $u_1 = u$ and s_t has to commute with u by the uniqueness of the one-parameter subgroup containing u. Thus for $g = su$ to be exponential it is necessary that there exists one-parameter subgroup $\{s_t\}$ containing s, and contained in the centralizer of u. It is easy to see that conversely if there exists such a one-parameter subgroup $\{s_t\}$ then g is exponential. This gives a convenient way of checking exponentiality of elements in these groups. In particular for most examples of Lie groups in § 2.3 that are asserted to be exponential it can be confirmed in this way to be the case. It is however not easy in general to put together the conditions for individual elements to get a criterion for the group to be exponential (in terms of group related data directly). A study of exponentiality of a class of groups using these ideas was carried out in [**6**], where an analogue of the result in the last bullet item in § 2.3 is proved for semidirect products of the form $G = K \cdot N$, where K is any compact connected Lie group and N is a simply connected nilpotent Lie group. We shall however not go into further details on this.

3.2. Conditions for exponentiality of all elements in a coset. Let G be a connected solvable Lie group and N be a closed normal simply connected Lie subgroup of positive dimension such that G/N is abelian. Let N_j, $j = 1, \dots k$, be closed connected normal subgroups of G contained in N such that $N_0 = N$, $N_k = \{e\}$, the trivial subgroup, and for all $j = 0, \dots, k-1$, N_{j+1} is proper subgroup of N_j containing $[N, N_j]$. For all $j = 1, \dots, k$ let $V_j = N_{j-1}/N_j$. The G-action on N by conjugation induces an action on V_j for all j. Moreover, the restriction of the action to N is trivial, and thus we get an action of G/N on each V_j for each $j = 1, \dots k$.

We shall now describe a criterion for all elements of a given coset xN, where $x \in G$, to be exponential. Let $A = G/N$. Let $x \in G$ be given and $a = xN \in G/N = A$. For all $j = 1, \dots, k$ let $V_j^a = \{v \in V_j \mid av = v\}$ and let

$$K = \{\alpha \in A \mid \alpha v = v \text{ for all } v \in V_j^a, j = 1, \dots, k\}.$$

We note that $a \in K$. We shall denote by K^0 the connected component of the identity in K. The following is a modified version of the main theorem in [**4**].

THEOREM 3.1. *Let the notation be as above. Then xn is exponential for all $n \in N$ if and only if $a \in K^0$. Moreover, if P is a one-parameter subgroup of G such that $xN \subset (PN/N) \subset K^0$ then for all $n \in N$, xn is contained in a one-parameter subgroup contained in PN.*

The assumption in the hypothesis of the theorem that N be simply connected is made purely for convenience of exposition. A general connected nilpotent Lie subgroup has a compact subgroup contained in the center modulo which it is simply connected. Using this the same conclusion as in the theorem can be upheld without the assumption of N being simply connected.

The statement in the last bullet item in § 2.3 can be readily deduced from this criterion, and similarly other criteria in literature in terms of the Cartan subgroups can be derived from it. We shall however not go into the details here.

3.3. Proof of Theorem 3.1 - a special case. In this section we shall prove a special case of the theorem. For introducing the condition involved we recall the following. The action of a group A on a (finite dimensional) vector space V is said to be *isotypical* if V can be decomposed as $V = \oplus_{i=1}^k V_i$, for some $k \in \mathbb{N}$, such that for each i, V_i is A-invariant, the action of A on V_i is irreducible, and the A-actions on all V_i, $i = 1, \dots, k$, are equivalent to each other (as group actions on vector spaces). We note that when the action of an abelian group A on a vector space V is isotypical, for any subgroup B of A on V is also isotypical (though the number of irreducible components could be more).

The special case sought to be proved first is under the condition that N is a vector space and the action of G on N (by conjugation as above) is isotypical, and hence so is the action of A. In this case we have $k = 1$ and $V_1 = N$. Since the A-action is isotypical, it follows that the action of a (viz. of the cyclic subgroup generated by a) is isotypical and hence V_1^a is either the whole of N or trivial. Suppose $V_1^a = N$. Then the action of K on N is trivial. Suppose $a \in K^0$. Then there exists a one-parameter subgroup of A containing a and contained in K^0, and it can be expressed as PN/N, with P a one-parameter subgroup of G. As $xN = a \in PN/N$, $x \in PN$. Since the K-action of N is trivial and PN/N is contained in K it follows that the P-action on N is trivial, which means that PN is an abelian (connected) Lie group. Therefore xn is exponential for all n. The additional statement in the theorem is also clear from this argument, in the case at hand.

Now consider the case when V_1^a is trivial. Then $K = A = K^0$ and there exist one-parameter subgroups in A containing a; let $\{a_t\}$ be one such, with $a_1 = a$. Let $P = \{g_t\}$ be a one-parameter subgroup of G such that $a_t = g_tN$ for all t. As $xN = a = a_1 = g_1N$, there exists $n_0 \in N$ such that $x = g_1n_0$. The action of g_1 on N, which is the same as that of a, has no nonzero fixed point, as V_1 is trivial. Noting that it is a linear action it is easy to deduce from the preceding observation that for every $v \in N$ there exists a (unique) $w \in N$ such that $g_1v = wg_1w^{-1}$. Let $n \in N$ be given; then choosing $v = n_0n$ we get that there exists $w \in N$ such that $xn = g_1n_0n = wg_1w^{-1}$. This shows that xn is contained in the one-parameter subgroup wPw^{-1}, contained in PN. Thus xn is exponential in PN. This proves the 'if' part and the refined statement in the theorem in the special case under consideration.

We now prove the converse statement by showing that if $a \notin K^0$ then there are at most countably many n in N such that xn is exponential. Let $n \in N$ be such that such that xn is exponential, and let $P = \{g_t\}$ be a one-parameter subgroup such that $g_1 = xn$. Let $a_t = g_tN$ for all t. Then $a_1 = a$ and since $a \notin K^0$ we get that $\{a_t\}$ is not contained in K. The P-action on N is same as the $\{a_t\}$-action. As $\{a_t\}$ is not contained in K there exists $v \in N$ which is fixed by a but not by all a_t, $t \in \mathbb{R}$. Since by assumption the A-action on N is isotypical, so is the $\{a_t\}$-action on N, and the preceding condition therefore implies that the $\{a_t\}$-action has no nontrivial fixed points on N. Thus the P-action has no nontrivial fixed point. It can be seen that when this holds, any one-parameter subgroups in PN which is not contained in N is of the form wPw^{-1}. Since g_1 commutes will all elements of N, this shows that there is no one-parameter subgroup in PN containing an element of the form g_1m, with $m \in N$ nontrivial. This shows that if n, n' are two distinct elements of N and xn and xn' are contained in one-parameter subgroups P and P' of G then PN and $P'N$ are distinct, and in turn PN/N and $P'N/N$ are distinct one-parameter subgroups of A, containing a. Since a can be contained in only countably many one-parameter subgroups (only one if A has no nontrivial compact subgroup) we get that when $a \notin K^0$ there can be only countably many n in N such that xn is exponential. This proves the converse, thus completing the proof of the theorem in the case at hand.

3.4. Proof of Theorem 3.1. We now consider the general case, following the notation set up earlier. Let $V = N_{k-1} = V_k$, which is a vector space of positive dimension; recall that on V we have an action of G, by conjugation, whose restriction to N is trivial and hence factors to an action of A. Let U be an A-invariant subspace of V of positive dimension on which the A-action is isotypical. Let $W = V/U$, equipped with the induced A-action. Let $U^a = \{u \in U \mid au = u\}$, $W^a = \{w \in W \mid aw = w\}$,

$$L_U = \{\alpha \in A \mid \alpha u = u, \ \forall\, u \in U^a\} \text{ and } L_W = \{\alpha \in A \mid \alpha w = w, \ \forall\, w \in W^a\}.$$

It is straightforward to verify using the Jordan canonical form of linear transformations that $\{\alpha \in A \mid \alpha v = v \text{ for all } v \in V_k^a\} = L_U \cap L_W$, (where V_k^a is as before).

Now let $G' = G/U$, $N' = N/U$, $x' = xU$, and $N_j' = N_j/U$ for all $j = 0, \dots k$. We note that G'/N' may be identified canonically with $A = G/N$. The observation above shows that if K' is the subgroup of A defined as in the formulation of the theorem for this data, in place of G, N, x and N_j, $j = 0, \dots, k$, then $K = K' \cap L_U$. Since G' is of dimension lower than G, by an obvious induction hypothesis we may assume that the conclusions of the theorem hold for this setup; the induction can be started either with trivial choices, or the special case proved in the preceding subsection. We shall now deduce the assertions as in the theorem, for G, N, x etc., from this.

Suppose that $a \in K^0$. Then there exists a one-parameter subgroup P of G such that $xN \subset (PN/N) \subset K^0$. As K is contained in K', by the induction hypothesis we get that for any $n' \in N'$, $x'n'$ is exponential in PN'. Now let $n \in N$ be given and let $n' = nU$. Then there exists a one-parameter subgroup P' of PN such that $x'n' \in P'U$, and hence $xn = pu$ for some $p \in P'$ and $u \in U$. We shall now apply the special case proved earlier to $P'U$ in place of G, and p in place of x. We note that since the A-action on U is isotypical and the P'-action is as a subgroup of A, the P' action is isotypical. Also since PN/N is contained in $K^0 \subset L_U$ and the p-action on U is the same as the action of a, it follows that every $u \in U$ which is fixed by p is fixed by P'. Hence from the special case we get that pu is exponential in $P'U$. Thus xn is exponential in $P'U \subset PN$. As this holds for all $n \in N$, this proves the 'if' part and also the refined assertion in theorem.

To prove the converse we proceed as follows. Let $\{N_j\}_{j=1}^k$ be the subgroups as in the formulation of the theorem. Let $\mathcal{Q}$ be the family of one-parameter subgroups of A containing a. We note that $\mathcal{Q}$ is a countable family; it has only one element if A has no compact subgroup of positive dimension. Suppose that $a \notin K^0$. For $Q \in \mathcal{Q}$ let $E(Q)$ denote the set of n in N such that xn is contained in a one-parameter subgroup P of G such that $Q = PN/N$. We shall show that for each Q there exists a j, $0 \leq j \leq k-1$, and a closed normal subgroup M of G such that $N_{j+1} \subset M \subsetneq N_j$ and the following holds: for $n \in E(Q)$ the set $\{m \in N_j \mid nm \in E(Q)\}$ is contained in M. The conclusion implies that $\cup_{Q \in \mathcal{Q}} E(Q)$ can not be the whole of N; one immediate way of seeing this would be to note that when this holds each $E(Q)$ has Haar measure 0 in N and use the countability of $\mathcal{Q}$ - but the reader can convince herself of the conclusion by other, geometric or algebraic, considerations. Since $\cup_{Q \in \mathcal{Q}} E(Q)$ must account for all the elements n in N such that xn is exponential this would prove the converse assertion in the theorem.

It remains to prove the claim as above for each Q. Let $Q \in \mathcal{Q}$ be given. Since $a \notin K^0$ we get that Q is not contained in K. Hence there exists a j, $0 \leq j \leq k-1$, for which there exists a $v \in V_{j+1}^a$ such that v is not fixed under the action of Q, and we choose j to be smallest integer for which this holds. Since there exists $v \in V_{j+1}$ which is fixed by a but not the whole of Q we can get a proper closed subgroup M of N_j containing N_{j+1},which is normal in G and such that the A-action on $U := N_j/M$ is isotypical and there exists a $u \in U$ which is fixed under the action a but not of Q. We now show that the above claim holds with respect to these choices. Let $n \in E(Q)$ be given. Thus we have a one-parameter subgroup P of N containing xn such that $Q = PN/N$. Consider the subgroup PU. We note that the P-action on U (by conjugation) is the same as the Q-action and that of xn coincides with the action of a. Then $u \in U$ as above fixed by the action of xn but not that of P. Also, since the A-action on U is isotypical the P-action on U is isotypical, and by the preceding conclusion it has no nonzero fixed point. Under this condition, as seen the proof in the special case there does not exist any nonzero w in U such that xnw is exponential in PU. Now if $m \in N_{j+1}$ is such that xnm is contained in a one-parameter subgroup of G which projects to Q modulo N, then mU must be trivial in U, namely $m \in M$. This proves the claim, and thus completes the proof of the theorem. $\square$

4. Complements

Let G be a connected Lie group. In place of the question of existence of roots of all orders for an element g, which by the theorem of McCrudden is equivalent to g being exponential, one may ask for roots of a given order, $k \geq 2$. Thus in analogy to the question of G being exponential (respectively weakly exponential) one can ask for the surjectivity of the power map P_k defined by $P_k(x) = x^k$ for all $x \in G$ (respectively the image of P_k being dense in G). These themes have been explored in papers of Pralay Chatterjee and more recently by Saurav Bhaumik and Arunava Mandal, in which various results analogous to the corresponding results for the exponential map are proved. An interested reader may consult [**3**], [**1**] and [**8**] in this respect. An interesting fallout that may be mentioned for its general interest is that for each connected semisimple Lie group G there exists an integer q (typically not very large) such that if a $g \in G$ admits roots of order q then it admits roots of all orders. Some of the theory in this respect has also been extended to algebraic groups in place of Lie groups.

A theorem analogous to Theorem 3.1 for individual power maps for a class of abstract Lie groups, including algebraic groups over various fields, is proved in [**5**], and is also applied to deduce a corollary similar to Corollary 2.2. We shall however not go into the details here.

References

[1] S. Bhaumik and A. Mandal, *On the density of images of the power maps in Lie groups*, Arch. Math. (Basel) **110** (2018), no. 2, 115–130, DOI 10.1007/s00013-017-1130-4. MR3746989

[2] P. Chatterjee, *On the surjectivity of the power maps of algebraic groups in characteristic zero*, Math. Res. Lett. **9** (2002), no. 5-6, 741–756, DOI 10.4310/MRL.2002.v9.n6.a4. MR1906075

[3] P. Chatterjee, *Surjectivity of power maps of real algebraic groups*, Adv. Math. **226** (2011), no. 6, 4639–4666, DOI 10.1016/j.aim.2010.11.005. MR2775880

[4] S.G. Dani, On the radicals of exponential Lie groups, Preprint (2016), available at `https://arxiv.org/pdf/1605.07758.pdf`

[5] S. G. Dani and A. Mandal, *On the surjectivity of the power maps of a class of solvable groups*, J. Group Theory **20** (2017), no. 6, 1089–1101, DOI 10.1515/jgth-2017-0013. MR3719318

[6] S. G. Dani and M. McCrudden, *A criterion for exponentiality in certain Lie groups*, J. Algebra **238** (2001), no. 1, 82–98, DOI 10.1006/jabr.2000.8642. MR1822184

[7] D. Ž. Đoković and K. H. Hofmann, *The surjectivity question for the exponential function of real Lie groups: A status report*, J. Lie Theory **7** (1997), no. 2, 171–199. MR1473164

[8] A. Mandal, *Dense images of the power maps in Lie groups and minimal parabolic subgroups*, New York J. Math. **24** (2018), 201–209. MR3778500

[9] M. McCrudden, *On nth roots and infinitely divisible elements in a connected Lie group*, Math. Proc. Cambridge Philos. Soc. **89** (1981), no. 2, 293–299, DOI 10.1017/S0305004100058175. MR600244

[10] M. Wüstner, *Supplements on the theory of exponential Lie groups*, J. Algebra **265** (2003), no. 1, 148–170, DOI 10.1016/S0021-8693(03)00165-0. MR1984904

[11] M. Wüstner, *The classification of all simple Lie groups with surjective exponential map*, J. Lie Theory **15** (2005), no. 1, 269–278. MR2115242

UM-DAE Center for Excellence in Basic Sciences,, University of Mumbai, Kalina, Mumbai 400098, India

Email address: `shrigodani@cbs.ac.in`

Contemporary Mathematics
Volume **738**, 2019
https://doi.org/10.1090/conm/738/14876

A note on Nullstellensatz over finite fields

Sudhir R. Ghorpade

Abstract. We give an expository account of Nullstellensatz-like results when the base field is finite. In particular, we discuss the vanishing ideal of the affine space and of the projective space over a finite field. As an application, we include an alternative proof of Ore's inequality for the number of points of affine hypersurfaces over finite fields.

1. Introduction

Hilbert's Nullstellensatz, or Hilbert's Zero Point Theorem, is a classical result of fundamental importance in commutative algebra and algebraic geometry. This result is only valid when the base field is $\mathbb{C}$, the field of complex numbers, or more generally an algebraically closed field. In fact, when the base field is $\mathbb{C}$, it can be viewed as a remarkable generalization of the Fundamental Theorem of Algebra. There are two versions, commonly known as Weak Nullstellensatz and Strong Nullstellensatz, which can be stated as follows.

Weak Nullstellensatz: Let k be an algebraically closed field. If I is a nonunit ideal of the polynomial ring $k[X_1, \dots, X_n]$, then I has a 'zero', i.e., there exists $(\alpha_1, \dots, \alpha_n) \in k^n$ such that $f(\alpha_1, \dots, \alpha_n) = 0$, for each $f \in I$.

This result is usually deduced from the assertion, also referred to as a Nullstellensatz, that every maximal ideal of $k[X_1, \dots, X_n]$ is of the form $(X_1 - \alpha_1, \dots, X_n - \alpha_n)$ for some $\alpha_1, \dots, \alpha_n \in k$, provided of course k is algebraically closed. On the other hand, the Weak Nullstellensatz together with a well-known "trick of Rabinowitsch", implies the following version.

Strong Nullstellensatz: Let k be an algebraically closed field and let $f_1, \dots, f_m$ be any polynomials in $k[X_1, \dots, X_n]$. If $f \in k[X_1, \dots, X_n]$ is such that f vanishes at every common zero in k^n of $f_1, \dots, f_m$, then

$$f^r = g_1 f_1 + \cdots + g_m f_m \quad \text{for some } g_1, \dots, g_m \in k[X_1, \dots, X_n] \text{ and } r \geq 0.$$

The above statement is close to the original version of the theorem as it appears in Hilbert's 1893 paper [**11**, §3] on the complete systems of invariants. Hilbert calls this a *third general theorem in the theory of algebraic functions, continuing*

2010 *Mathematics Subject Classification.* 14G15, 11T06, 13B25, 11G25, 14G05.

Key words and phrases. Nullstellensatz, finite field, vanishing ideal, affine space, projective space, footprint.

Partially supported by the DST-RCN grant INT/NOR/RCN/ICT/P-03/2018 from the Dept. of Science & Technology, Govt. of India, MATRICS grant MTR/2018/000369 from the Science and Engg. Research Board, Govt. of India, and IRCC Award grant 12IRAWD009 from IIT Bombay.

Theorems I and III of his 1890 paper [**10**, §I, III] on the theory of algebraic forms. These latter theorems being what we now call Hilbert's basis theorem and Hilbert's syzygy theorem. A translation into English of Hilbert's papers on invariant theory is now available (cf. [**12**]) and one can access more easily the writings of a master. Nowadays, (Strong) Nullstellensatz is more commonly stated in the language of vanishing ideals of affine algebraic varieties and we recall this version in § 2.1 below. We also recall analogous result for projective varieties that one calls Projective Nullstellensatz. Most modern textbooks on commutative algebra contain a proof of Hilbert's Nullstellensatz, and we refer to Eisenbud's book [**8**] which has five different proofs, and to [**1, 17, 19, 24**] for a sampling of alternative proofs. See also the article by Goel, Patil and Verma [**9**] in this volume and the older article by Laksov [**15**] where some variations of Hilbert's Nullstellensatz are discussed.

A trivial consequence of Strong Nullstellensatz is that if k is an algebraically closed field and if $f \in k[X_1, \dots, X_n]$ vanishes on all of k^n, then f is the zero polynomial. We will refer to it as **Very Weak Nullstellensatz**, or in short, **VWN**. To deduce it from Strong Nullstellensatz, it suffices to take $m = 1$ and f_1 to be the zero polynomial. The VWN is, in fact, valid if k is any infinite field, and can be proved easily using induction on n and noting that a polynomial in one variable of degree d with coefficients in a field k has at most d roots in k.

On the other hand, even the VWN is not true if the base field is finite, say the finite field $\mathbb{F}_q$ with q elements. Indeed, there are nonzero polynomials such as $X_i^q - X_i$ that vanish on all of $\mathbb{F}_q^n$. Nonetheless one has the following result, which may be viewed as an analogue of (Very Weak) Nullstellensatz over finite fields.

Affine $\mathbb{F}_q$-Nullstellensatz: Let $f \in \mathbb{F}_q[X_1, \dots, X_n]$ and let Γ_q denote the ideal of $\mathbb{F}_q[X_1, \dots, X_n]$ generated by $X_1^q - X_1, \dots, X_n^q - X_n$. Then:

(i) f vanishes at every point of $\mathbb{F}_q^n$ if and only if $f \in \Gamma_q$.

(ii) Let $f_1, \dots, f_m$ be polynomials in $\mathbb{F}_q[X_1, \dots, X_n]$. If f vanishes at every common zero of $f_1, \dots, f_m$ in $\mathbb{F}_q^n$, then

$$f = g_1 f_1 + \cdots + g_m f_m + \gamma \quad \text{for some } g_1, \dots, g_m \in \mathbb{F}_q[X_1, \dots, X_n] \text{ and } \gamma \in \Gamma_q.$$

Note that (i) is a special case of (ii) and also that in (ii), one doesn't have to take a power of f (unlike in Strong Nullstellensatz). In other words, if I is an ideal of $\mathbb{F}_q[X_1, \dots, X_n]$, then $I + \Gamma_q$ is automatically a radical ideal of $\mathbb{F}_q[X_1, \dots, X_n]$. The above result is not new and goes back at least to Terjanian [**23**]. An excellent account is available in the article (in French) of Joly [**13**]; see also Delsarte, Goethals, and MacWilliams[**7**, § 1]. A more modern reference is Kreuzer and Robbiano [**14**, § 6.2A]. However, in the experience of the author, the result is not as widely known as it should be. Moreover, an analogue of (i) in the projective case that gives an explicit description of homogeneous polynomials that vanish on all of $\mathbb{P}^n(\mathbb{F}_q)$, appears to be known to even fewer algebraists. This is of a relatively recent vintage and may be attributed to Mercier and Rolland [**18**] (see also Remark 3.4). We thus provide in this article a self-contained account of these Nullstellensatz-like results in the setting of finite fields. Our proof of the projective analogue of (i) above uses the notion of projective reduction developed in [**3**] and is a little simpler than the original proof of Mercier and Rolland. We will also point out that a straightforward analgoue of (ii) in the projective case is not possible. In the affine case, the Affine $\mathbb{F}_q$-Nullstellensatz can be useful to deduce the so called affine $\mathbb{F}_q$-footprint bound for estimating the number of $\mathbb{F}_q$-rational points of affine algebraic varieties defined over $\mathbb{F}_q$. We outline this and show how such a bound can be used to deduce a classical inequality for the number of points of affine hypersurfaces defined over $\mathbb{F}_q$.

As indicated in the abstract, this is an expository article, and we have made an attempt to keep it fairly self-contained. The results given here are not new, but are somewhat scattered in the literature not all of which is easily available in English. The proofs given here of some of the results (especially in Section 3) appear to be simpler and more natural than those available elsewhere in the literature.

2. The Affine Case

In the first subsection, we set up some basic notation and recall preliminaries about affine varieties. The notion of reduced polynomials is discussed in the next subsection and a couple of auxiliary results are proved here. These are then used in § 2.3 to prove the result described in the Introduction as the Affine $\mathbb{F}_q$- Nullstellensatz, and, in fact, a slightly more general version of it. Finally, in § 2.4, we give an application to a classical inequality due to Ore. Our exposition in subsections 2.2 and 2.3 closely follows Joly [**13**, Ch. 2], while the proof of Lemma 2.6 in § 2.3 is adapated from Carvalho's notes [**4**, § 3].

2.1. Preliminaries. Let k be a field and let n be a nonnegative integer. Also, let $S := k[X_1, \dots, X_n]$ be the ring of polynomials in n variables $X_1, \dots, X_n$ with coefficients in k. We denote by $\mathbb{A}_k^n$ (or simply, $\mathbb{A}^n$ if the reference to k is clear from the context) the space of n-tuples of elements of k. Given any $I \subseteq S$, we let

$$\mathsf{Z}(I) := \{\mathbf{a} = (a_1, \dots, a_n) \in \mathbb{A}_k^n : f(a_1, \dots, a_n) = 0 \text{ for all } f \in I\}.$$

Given a subfield F of k, we call a subset Z of $\mathbb{A}_k^n$ an *affine algebraic variety* defined over F if $Z = \mathsf{Z}(I)$ for some $I \subseteq F[X_1, \dots, X_n]$. This is equivalent to saying that $Z = \mathsf{Z}(I)$ for some ideal I of S generated by finitely many polynomials in $F[X_1, \dots, X_n]$. If Z is an affine algebraic variety defined over F and if K is a field extension of F such that K is a subfield of an algebraic closure $\overline{k}$ of k (so that F, K and k are subfields of $\overline{k}$), then we denote by $Z(K)$ the set of K-rational points of Z, i.e., $Z(K) := \{\mathbf{a} \in \mathbb{A}_K^n : f(\mathbf{a}) = 0 \text{ for all } f \in I\}$. Given a subset Z of $\mathbb{A}_k^n$, we let

$$\mathsf{I}(Z) := \{f \in S : f(\mathbf{a}) = 0 \text{ for all } \mathbf{a} \in Z\},$$

and we note that $\mathsf{I}(Z)$ is an ideal of S; it is called the *vanishing ideal* of Z. It is not difficult to see that if Z is an affine algebraic variety defined over k, then $\mathsf{Z}(\mathsf{I}(Z)) = Z$. On the other hand, Strong Nullstellensatz can be stated as follows.

$$\mathsf{I}(\mathsf{Z}(I)) = \sqrt{I} \quad \text{if } k \text{ is algebraically closed and } I \text{ is any ideal of } S. \tag{2.1}$$

By considering an ideal I of S such that $\mathsf{Z}(I) = \emptyset$, we obtain the Weak Nullstellensatz, whereas by considering the special case $\mathsf{Z}(I) = \mathbb{A}_k^n$, we can deduce the VWN. As noted in the Introduction, the latter is valid (and rather easily proved) more generally when k is an infinite field that is not necessarily algebraically closed.

2.2. Reduction. Fix a finite field $\mathbb{F}_q$ with q elements (so that q is a prime power) and an algebraic closure $\overline{\mathbb{F}}_q$ of $\mathbb{F}_q$ (in fact, we can take $\overline{\mathbb{F}}_q$ to be $\cup_{n\geq 1}\mathbb{F}_{q^n}$). Let k be a subfield of $\overline{\mathbb{F}}_q$ containing $\mathbb{F}_q$ (or equivalently, k is an algebraic extension of $\mathbb{F}_q$). As before, let $S := k[X_1, \dots, X_n]$.

A polynomial $f \in S$ is said to be *reduced* if $\deg_{X_i} f \leq q-1$ for each $i = 1, \dots, n$. The set of all reduced polynomials in S will be denoted by $\mathfrak{R}$. Clearly, $\mathfrak{R}$ is a vector space over k and the monomials $X_1^{i_1} \cdots X_n^{i_n}$, where $0 \le i_j \le q-1$ for $j = 1, \dots, n$, form a k-basis of $\mathfrak{R}$. In particular, $\dim \mathfrak{R} = q^n$. Note that if $n = 0$, then $\mathfrak{R} = S = k$.

LEMMA 2.1. *If $f \in \mathfrak{R}$ is such that $f(\mathbf{a}) = 0$ for all $\mathbf{a} \in \mathbb{A}^n(\mathbb{F}_q)$, then f is the zero polynomial. In other words, $\mathfrak{R} \cap \mathsf{I}(\mathbb{A}^n(\mathbb{F}_q)) = \{0\}$.*

PROOF. The case when $n = 0$ is trivial. Suppose $n > 0$ and the result holds for polynomials in $n-1$ variables. Let $f \in \mathfrak{R}$ be such that $f(\mathbf{a}) = 0$ for all $\mathbf{a} \in \mathbb{A}^n(\mathbb{F}_q)$. Since f is reduced, we can write $f = f(X_1, \dots, X_n)$ as

$$f = f_0 X_n^{q-1} + f_1 X_n^{q-2} + \cdots + f_{q-1} \quad \text{where } f_i \in k[X_1, \dots, X_{n-1}] \text{ for } i = 0, 1, \dots, q-1.$$

Now for any fixed $(a_1, \dots, a_{n-1}) \in \mathbb{A}^{n-1}(\mathbb{F}_q)$, the polynomial $f(a_1, \dots, a_{n-1}, X_n)$ in $k[X_n]$ has degree $\le q-1$ and has at least q roots in k. Hence it must be the zero polynomial. Consequently, $f_i(a_1, \dots, a_{n-1}) = 0$ for all $i = 0, 1, \dots, q-1$ and $(a_1, \dots, a_{n-1}) \in \mathbb{A}^{n-1}(\mathbb{F}_q)$. So by the induction hypothesis, each f_i is the zero polynomial, and therefore, so is f. □

Now let us define $\Gamma_q(k)$ to be the ideal of $S = k[X_1, \dots, X_n]$ generated by $X_1^q - X_1, \dots, X_n^q - X_n$. Clearly, $\Gamma_q(k) \subseteq \mathsf{I}(\mathbb{A}^n(\mathbb{F}_q))$, and so Lemma 2.1 implies that

$$\mathfrak{R} \cap \Gamma_q(k) = \{0\}. \tag{2.2}$$

LEMMA 2.2. *Every $f \in S$ can be uniquely written as $f = g + \gamma$ for some $g \in \mathfrak{R}$ and $\gamma \in \Gamma_q(k)$. In other words, $S = \mathfrak{R} \oplus \Gamma_q(k)$.*

PROOF. The uniqueness is clear from (2.2) since both $\mathfrak{R}$ and $\Gamma_q(k)$ are clearly vector spaces over k. To prove the existence, it suffices to take f to be a monomial, say $f = X_1^{i_1} \cdots X_n^{i_n}$. If f is not reduced, then $i_j \ge q$ for some $j \in \{1, \dots, n\}$. Note that

$$X_j^{i_j} = X_j^{i_j - q}(X_j^q - X_j + X_j) \equiv X_j^{i_j-(q-1)} (\text{mod } \Gamma_q(k)).$$

Hence $f \equiv X_1^{i_1} \cdots X_{j-1}^{i_{j-1}} X_j^{i_j-(q-1)} X_{j+1}^{i_{j+1}} \cdots X_n^{i_n} (\text{mod } \Gamma_q(k))$. Continuing in this way, we see that $f \equiv g (\text{mod } \Gamma_q(k))$ for some reduced monomial g. □

2.3. Affine $\mathbb{F}_q$-Nullstellensatz. We will continue to use the notation and terminology in § 2.1 and § 2.2. The Affine $\mathbb{F}_q$-Nullstellensatz stated in the Introduction is a special case of the theorem below with $k = \mathbb{F}_q$, where $Z(\mathbb{F}_q)$ coincides with $Z(I)$.

THEOREM 2.3. *Let k be an algebraic extension of $\mathbb{F}_q$. Then:*

(i) *$\mathsf{I}(\mathbb{A}^n(\mathbb{F}_q)) = \Gamma_q(k)$.*

(ii) *If Z is an affine algebraic variety in $\mathbb{A}_k^n$ defined over $\mathbb{F}_q$ and $Z = \mathsf{Z}(I)$ for some ideal I of S generated by polynomials in $\mathbb{F}_q[X_1, \dots, X_n]$, then*

$$\mathsf{I}(Z(\mathbb{F}_q)) = I + \Gamma_q(k).$$

PROOF. (i) The inclusion $\Gamma_q(k) \subseteq \mathsf{I}(\mathbb{A}^n(\mathbb{F}_q))$ is obvious. To prove the reverse inclusion, suppose $f \in \mathsf{I}(\mathbb{A}^n(\mathbb{F}_q))$. By Lemma 2.2, we can write $f = g + \gamma$ for some $g \in \mathfrak{R}$ and $\gamma \in \Gamma_q(k)$. But then $g = f - \gamma$ vanishes on $\mathbb{A}^n(\mathbb{F}_q)$ and so by Lemma 2.1, $g = 0$. Thus $f \in \Gamma_q(k)$.

(ii) Let I be an ideal of S generated by polynomials in $\mathbb{F}_q[X_1, \dots, X_n]$ and let $Z = \mathsf{Z}(I)$. Evidently, $I + \Gamma_q(k) \subseteq \mathsf{I}(Z(\mathbb{F}_q))$. To prove the reverse inclusion, first note that by Hilbert basis theorem, $I = \langle f_1, \dots, f_r \rangle$ for some $f_1, \dots, f_r \in \mathbb{F}_q[X_1, \dots, X_n]$. Let us consider

$$g := 1 - \left(1 - f_1^{q-1}\right) \cdots \left(1 - f_r^{q-1}\right) \quad \text{and} \quad h := 1 - g.$$

It is clear that $g, h \in \mathbb{F}_q[X_1, \dots, X_n]$ and $g \in I$. Moreover,

$$g(\mathbf{a}) = \begin{cases} 0 & \text{if } \mathbf{a} \in Z(\mathbb{F}_q) \\ 1 & \text{if } \mathbf{a} \in \mathbb{A}^n(\mathbb{F}_q) \setminus Z(\mathbb{F}_q) \end{cases} \quad \text{and} \quad h(\mathbf{a}) = \begin{cases} 1 & \text{if } \mathbf{a} \in Z(\mathbb{F}_q) \\ 0 & \text{if } \mathbf{a} \in \mathbb{A}^n(\mathbb{F}_q) \setminus Z(\mathbb{F}_q). \end{cases}$$

Now let $f \in \mathsf{I}(Z(\mathbb{F}_q))$. Then $fg \in I$ and $fh \in \mathsf{I}(\mathbb{A}^n(\mathbb{F}_q))$. Since $g + h = 1$, we see from (i) above that $f = fg + fh \in I + \Gamma_q(k)$. Thus $\mathsf{I}(Z(\mathbb{F}_q)) = I + \Gamma_q(k)$. □

REMARK 2.4. An immediate corollary of part (ii) of the above theorem is that for any ideal I of S generated by polynomials in $\mathbb{F}_q[X_1, \dots, X_n]$, the ideal $I + \Gamma_q(k)$ is a radical ideal. This particular fact can also be deduced from Seidenberg's Lemma given in Article 92 of [**21**].

2.4. Application to Affine Hypersurfaces over Finite Fields. Let k be an arbitrary field and as before, let S denote the polynomial ring $k[X_1, \dots, X_n]$. Also, let $\mathcal{M}$ denote the set of all monomials in S (including the constant monomial 1). Fix a monomial order on $\mathcal{M}$, i.e., a total order $\preccurlyeq$ on $\mathcal{M}$ satisfying (i) $1 \preccurlyeq \mu$ for all $\mu \in \mathcal{M}$ and (ii) $\mu_1 \preccurlyeq \mu_2 \Rightarrow \nu\mu_1 \preccurlyeq \nu\mu_2$ for all $\mu_1, \mu_2, \nu \in \mathcal{M}$. For $0 \neq f \in S$, let $\text{in}_{\preccurlyeq}(f)$ denote the largest monomial (w.r.t. $\preccurlyeq$) appearing in f with a nonzero coefficient; this is called the *leading monomial* or the *initial monomial* of f (w.r.t. $\preccurlyeq$). For any subset I of S, define the *footprint* of I to be the set $\Delta(I)$ of all monomials in $\mathcal{M}$ that are not divisible by the leading monomials of any nonzero element of I, i.e.,

$$\Delta(I) := \{\mu \in \mathcal{M} : \text{in}_{\preccurlyeq}(f) \nmid \mu \text{ for all } f \in I \text{ with } f \neq 0\}.$$

If $I \subseteq S$ is finite, say $I = \{f_1, \dots, f_r\}$, then we may write $\Delta(I)$ as $\Delta(f_1, \dots, f_r)$.

The following result due to Buchberger is classical and is easily derived from the division algorithm (w.r.t. $\preccurlyeq$) in S. See, for example, Prop. 1 in Ch. 5, § 3 of [**5**].

PROPOSITION 2.5. $\{\mu + I : \mu \in \Delta(I)\}$ is a k-basis of S/I for any ideal I of S.

We can use this and a variant of Lagrange interpolation to derive a useful bound for the number of $\mathbb{F}_q$-rational points of affine algebraic varieties defined over $\mathbb{F}_q$.

LEMMA 2.6 (Affine $\mathbb{F}_q$-Footprint Bound). *Let k be an algebraic extension of $\mathbb{F}_q$ and let I be an ideal of $S := k[X_1, \dots, X_n]$ generated by some nonzero polynomials $f_1, \dots, f_r \in \mathbb{F}_q[X_1, \dots, X_n]$. Also let $Z = \mathsf{Z}(I)$ denote the corresponding affine algebraic variety in $\mathbb{A}^n_k$ defined over $\mathbb{F}_q$. Then*

$$|Z(\mathbb{F}_q)| \leq \left|\overline{\Delta}(f_1, \dots, f_r)\right|,$$

where $\overline{\Delta}(f_1, \dots, f_r) := \{\mu \in \mathcal{M} : \mu$ is reduced and $\text{in}_{\preccurlyeq}(f_i) \nmid \mu$ for $i = 1, \dots, r\}$.

PROOF. If $Z(\mathbb{F}_q)$ is empty, then there is nothing to prove. Thus, assume that $Z(\mathbb{F}_q)$ is nonempty. Let $I_q := I + \Gamma_q(k)$. Clearly, $Z(I_q) = Z(\mathbb{F}_q)$. Hence $Z(I_q)$ is a nonempty finite subset of $\mathbb{A}^n_k$; in particular, $I_q \neq S$. Write $Z(I_q) = \{\mathbf{a}_1, \dots, \mathbf{a}_m\}$, where $\mathbf{a}_i \neq \mathbf{a}_j$ for $1 \leq i < j \leq m$. Write $\mathbf{a}_i = (a_{i1}, \dots, a_{in})$ for $i = 1, \dots, m$. Fix $i \in \{1, \dots, m\}$. Then for each $j \in \{1, \dots, m\}$ with $j \neq i$, there is $t_j \in \{1, \dots, n\}$ such that $a_{it_j} \neq a_{jt_j}$. Consider

$$p_i(X_1, \dots, X_n) = \prod_{\substack{1 \leq j \leq m \\ j \neq i}} \frac{X_{t_j} - a_{jt_j}}{a_{it_j} - a_{jt_j}}.$$

The polynomials $p_1, \ldots, p_m \in \mathbb{F}_q[X_1, \ldots, X_n]$ thus obtained have the property that $p_i(\mathbf{a}_j) = \delta_{ij}$ for $i, j = 1, \ldots, m$. Moreover, $I_q = I(Z(\mathbb{F}_q))$, by part (ii) of Theorem 2.3. Hence if $\sum_{j=1}^m \lambda_j p_j \in I_q$ for some $\lambda_1, \ldots, \lambda_m \in k$, then by evaluating at $\mathbf{a}_i$, we obtain $\lambda_i = 0$ for $i = 1, \ldots, m$. It follows that $\{p_1 + I_q, \ldots, p_m + I_q\}$ is a k-linearly independent subset of S/I_q. Thus, by Proposition 2.5, we see that

$$|Z(\mathbb{F}_q)| = m \leq \dim_k S/I_q = |\Delta(I_q)| \leq |\Delta(f_1, \ldots, f_r, X_1^q - X_1, \ldots, X_n^q - X_n)|.$$

Finally, since $\text{in}_{\preccurlyeq}(X_i^q - X_i) = X_i^q$ for each $i = 1, \ldots, n$, it is clear that $\mu \in \mathcal{M}$ is reduced if and only if $\text{in}_{\preccurlyeq}(X_i^q - X_i) \nmid \mu$ for all $i = 1, \ldots, n$. This readily implies that $\Delta(f_1, \ldots, f_r, X_1^q - X_1, \ldots, X_n^q - X_n) = \overline{\Delta}(f_1, \ldots, f_r)$. □

As a corollary, we shall deduce an inequality that according to [**16**, p. 320], goes back at least to Ore (1922) and provides an effective bound on the number of $\mathbb{F}_q$-rational points of an affine hypersurface defined over $\mathbb{F}_q$ in terms of its degree. It can also be viewed as a multivariable generalization of the elementary fact that a univariate polynomial of degree d with coefficients in a field has at most d roots in that field. The generalization is of course possible when the base field is finite.

COROLLARY 2.7 (Ore's Inequality). *Let $f \in \mathbb{F}_q[X_1, \ldots, X_n]$ be a nonzero polynomial of degree d and let $Z = Z(f)$ be the corresponding variety in $\mathbb{A}_k^n$, where k is any algebraic extension of $\mathbb{F}_q$. Then*

$$|Z(\mathbb{F}_q)| \leq dq^{n-1}.$$

PROOF. The inequality is trivial if $d \geq q$ because then $dq^{n-1} \geq |\mathbb{A}^n(\mathbb{F}_q)| \geq |Z(\mathbb{F}_q)|$. Assume that $d < q$. This implies, in particular, that f is reduced. Fix a monomial order $\preccurlyeq$ on the set $\mathcal{M}$ of all monomials in $S := k[X_1, \ldots, X_n]$, and write $\text{in}_{\preccurlyeq}(f) = X_1^{i_1} \cdots X_n^{i_n}$, where $i_1, \ldots, i_n$ are nonnegative integers such that $i_1 + \cdots + i_n \leq d$. Note that $0 \leq i_j \leq d \leq q - 1$ for $j = 1, \ldots, n$. By Lemma 2.6,

$$|Z(\mathbb{F}_q)| \leq |\overline{\Delta}(f)| = \left|\{\mu \in \mathcal{M} : \mu \text{ is reduced and } X_1^{i_1} \cdots X_n^{i_n} \nmid \mu\}\right|.$$

Now a monomial $\mu = X_1^{j_1} \cdots X_n^{j_n}$ is reduced and is divisible by $X_1^{i_1} \cdots X_n^{i_n}$ if and only if $i_t \leq j_t \leq q - 1$ for all $t = 1, \ldots, n$. The number of such monomials is therefore $(q - i_1) \cdots (q - i_n)$. An easy induction on n shows that

$$(q - i_1) \cdots (q - i_n) \geq q^n - (i_1 + \cdots + i_n)q^{n-1} \qquad \text{whenever } 0 \leq i_j < q \text{ for } j = 1, \ldots, n.$$

Since the number of reduced monomials in $\mathcal{M}$ is clearly q^n, it follows that

$$|Z(\mathbb{F}_q)| \leq |\overline{\Delta}(f)| \leq q^n - \left(q^n - dq^{n-1}\right) = dq^{n-1},$$

where we have used the fact that $i_1 + \cdots + i_n \leq d$. □

3. Projective Version

In this section, we will try to develop projective analogues of the results in the first three subsections of Section 2. Our treatment will, in fact, be very parallel to that in §2.1-2.3.

3.1. Preliminaries. Let k be a field. The projective n-space over k will be denoted by $\mathbb{P}_k^n$ and it is the set of equivalence classes of elements of the set $k^{n+1} \setminus \{\mathbf{0}\}$ w.r.t. the equivalence relation $\sim$ given by proportionality, i.e., for all $\mathbf{a} = (a_0, a_1, \ldots, a_n)$ and $\mathbf{b} = (b_0, b_1, \ldots, b_n)$ in $k^{n+1} \setminus \{\mathbf{0}\}$,

$$\mathbf{a} \sim \mathbf{b} \iff \text{there is } \lambda \in k \text{ such that } a_i = \lambda b_i \text{ for all } i = 0, 1, \ldots, n.$$

We may denote by $[a_0 : a_1 : \cdots : a_n]$ the equivalence class of $(a_0, a_1, \ldots, a_n)$ in $k^{n+1} \setminus \{\mathbf{0}\}$. We let $\mathcal{S}$ be the polynomial ring $k[X_0, X_1, \ldots, X_n]$ in $n+1$ variables. Given a subset $\mathcal{I}$ of $\mathcal{S}$ consisting of homogeneous polynomials, we let

$$\mathsf{V}(\mathcal{I}) := \{[a_0 : a_1 : \cdots : a_n] \in \mathbb{P}_k^n : f(a_0, a_1, \ldots, a_n) = 0 \text{ for all } f \in \mathcal{I}\}.$$

Given a subfield F of k, we call a subset V of $\mathbb{P}_k^n$ a *projective algebraic variety* defined over F if $V = \mathsf{V}(\mathcal{I})$ for some subset $\mathcal{I}$ of homogeneous polynomials in $F[X_0, X_1, \ldots, X_n]$. This is equivalent to saying that $V = \mathsf{V}(\mathcal{I})$ for some homogeneous ideal $\mathcal{I}$ of S generated by finitely many homogeneous polynomials in $F[X_0, X_1, \ldots, X_n]$. If V a projective algebraic variety defined over F and if K is a field extension of F such that K is a subfield of an algebraic closure $\overline{k}$ of k, then we denote by $V(K)$ the set of K-rational points of V, i.e.,

$$V(K) := \{[a_0 : \cdots : a_n] \in \mathbb{P}_K^n : f(a_0, \ldots, a_n) = 0 \text{ for all homogeneous } f \in \mathcal{I}\}.$$

Given a subset V of $\mathbb{P}_k^n$, the *vanishing ideal* of V is defined to be the ideal $\mathsf{I}(V)$ of $\mathcal{S}$ generated by the homogeneous polynomials in $\mathcal{S}$ that vanish at every point of V. It is not difficult to see that if V is any projective algebraic variety defined over k, then $\mathsf{V}(\mathsf{I}(V)) = V$, while the projective analogue of (2.1) is the following.

Projective Nullstellensatz: If k is algebraically closed and if $\mathcal{I}$ is any homogeneous ideal of $\mathcal{S}$, then $\sqrt{\mathcal{I}} \supseteq \langle X_0, X_1, \ldots, X_n \rangle$ if $\mathsf{V}(\mathcal{I})$ is empty, whereas $\mathsf{I}(\mathsf{V}(\mathcal{I})) = \sqrt{\mathcal{I}}$ if $\mathsf{V}(\mathcal{I})$ is nonempty.

In particular, we see that there are nonunit homogeneous ideals of $\mathcal{S}$ that have no 'zero' in $\mathbb{P}_k^n$, even when k is algebraically closed. Thus a straightforward analogue of the Weak Nullstellensatz isn't quite true for projective varieties. On the other hand, the other special case $\mathsf{V}(\mathcal{I}) = \mathbb{P}_k^n$ still yields the projective analogue of VWN, which says that if k is algebraically closed, then the only homogeneous polynomial that vanishes on all of $\mathbb{P}_k^n$ is the zero polynomial. As before, this is valid more generally (and proved rather easily) when k is any infinite field. But when k is a finite field, such a result is not true and we discuss next what happens in this case.

3.2. Projective reduction. We follow [**3**] to outline here a projective analogue of the notion of reduction that was discussed in § 2.2. In this section, let k be an algebraic extension of $\mathbb{F}_q$ and let $\mathcal{S} := k[X_0, X_1, \ldots, X_n]$. Given a nonnegative integer d, let $\mathcal{S}_d$ denote the set of homogeneous polynomials in $\mathcal{S}$ of degree d (including the zero polynomial). We will denote by $\mathbb{M}$ the set of all monomials in $\mathcal{S}$. A monomial $\mu \in \mathbb{M}$ is said to be *projectively reduced* if either $\mu = 1$ or $\mu \neq 1$ and $\deg_{X_i} \mu \leq q-1$ for $1 \leq i < \ell_\mu$, where ℓ_μ is the index of the last variable appearing in μ, i.e., $\ell_\mu := \max\{\ell \in \{0, 1, \ldots, n\} : X_\ell \mid \mu\}$. Given a nonnegative integer d, we let $\mathfrak{R}_d$ denote the set of k-linear combinations of projectively reduced monomials in $\mathbb{M}$ of degree d. Clearly $\mathfrak{R}_d$ is a finite dimensional vector space over k and its elements may be called projectively reduced homogeneous polynomials of degree d. The next two results are projective analogues of Lemmas 2.1 and 2.2.

LEMMA 3.1. *Let d be a nonnegative integer. If $f \in \mathfrak{R}_d$ is such that $f(P) = 0$ for all $P \in \mathbb{P}^n(\mathbb{F}_q)$, then f is the zero polynomial. In other words, $\mathfrak{R}_d \cap \mathsf{I}(\mathbb{P}^n(\mathbb{F}_q)) = \{0\}$.*

PROOF. The case when $n = 0$ is trivial. Suppose $n > 0$ and the result holds for homogeneous polynomials of degree d in $k[X_0, X_1, \ldots, X_{n-1}]$. Let $f \in \mathfrak{R}_d \cap \mathsf{I}(\mathbb{P}^n(\mathbb{F}_q))$. By separating terms involving X_n, we can write

$$f(X_1, \ldots, X_n) = g(X_0, \ldots, X_{n-1}) + h(X_0, \ldots, X_n)$$

where $g \in k[X_0, \ldots, X_{n-1}]$ and $h \in k[X_0, \ldots, X_n]$ are homogeneous of degree d such that the last variable in each of the monomials appearing in h (with a nonzero coefficient) is X_n. Considering points $[a_0 : \cdots : a_n]$ of $\mathbb{P}^n(\mathbb{F}_q)$ such that $a_n = 0$, we deduce from the induction hypothesis that $g(X_0, \ldots, X_{n-1})$ is the zero polynomial. On the other hand, the dehomogenization $h(X_0, \ldots, X_{n-1}, 1)$ is a reduced (and not necessarily homogeneous) polynomial in $k[X_0, \ldots, X_{n-1}]$ that vanishes on $\mathbb{A}^n(\mathbb{F}_q)$. Hence by Lemma 2.1, $h(X_0, \ldots, X_{n-1}, 1)$ is the zero polynomial. Since h is a homogeneous polynomial divisible by X_n, it follows that h is also the zero polynomial. □

Now let us define $\Gamma_q^*(k)$ to be the ideal of $\mathcal{S}$ generated by the $\binom{n+1}{2}$ *Fermat polynomials* $X_i^q X_j - X_i X_j^q$, where $0 \le i < j \le n$. Clearly, $\Gamma_q^*(k)$ is a homogeneous ideal of $\mathcal{S}$ and $\Gamma_q^*(k) \subseteq \mathsf{I}(\mathbb{P}^n(\mathbb{F}_q))$. For any $d \ge 0$, we let $\Gamma_q^*(k)_d := \Gamma_q^*(k) \cap \mathcal{S}_d$. From Lemma 3.1, we see that

$$\mathfrak{R}_d \cap \Gamma_q^*(k)_d = \{0\} \quad \text{for every nonnegative integer } d. \tag{3.1}$$

LEMMA 3.2. *Let d be a nonnegative integer and let $f \in \mathcal{S}_d$. Then $f = g + \gamma$ for unique $g \in \mathfrak{R}_d$ and $\gamma \in \Gamma_q^*(k)_d$. Consequently, $\mathcal{S}_d = \mathfrak{R}_d \oplus \Gamma_q^*(k)_d$.*

PROOF. The uniqueness is clear from (3.1) since both $\mathfrak{R}_d$ and $\Gamma_q^*(k)_d$ are clearly vector spaces over k. To prove the existence, it suffices to take f to be a nonconstant monomial of degree d, say $f = X_0^{i_0} \cdots X_\ell^{i_\ell}$, where ℓ is the index of the last variable in f so that $i_\ell > 0$. If f is not projectively reduced, then $\ell \ge 1$ and $i_j \ge q$ for some $j \in \mathbb{Z}$ with $0 \le j < \ell$. Now observe that $X_j^{i_j} X_\ell^{i_\ell}$ can be written as

$$X_j^{i_j - q} X_\ell^{i_\ell - 1} \left(X_j^q X_\ell - X_j X_\ell^q + X_j X_\ell^q \right) \equiv X_j^{i_j - (q-1)} X_\ell^{i_\ell + (q-1)} (\text{mod } \Gamma_q^*(k)).$$

This implies that

$$f \equiv X_1^{i_1} \cdots X_{j-1}^{i_{j-1}} X_j^{i_j - (q-1)} X_{j+1}^{i_{j+1}} \cdots X_\ell^{i_\ell + (q-1)} (\text{mod } \Gamma_q^*(k)).$$

Continuing in this way, we see that $f \equiv g (\text{mod } \Gamma_q^*(k))$ for some projectively reduced monomial g of degree d. Moreover, $\gamma := f - g$ is necessarily a homogeneous polynomial in $\Gamma_q^*(k)$ of degree d. □

3.3. Vanishing Ideal of Projective Spaces over Finite Fields. We will continue to use the notation and terminology in § 3.1 and § 3.2. The following result is a slightly more general version of a theorem of Mercier and Rolland [**18**, Thm. 2.1]; in fact, it corresponds precisely to [**3**, Cor. 2.6]. The proof, however, is different from that in [**18**] or [**3**].

THEOREM 3.3. *Let k be an algebraic extension of $\mathbb{F}_q$. Then*

$$\mathsf{I}(\mathbb{P}^n(\mathbb{F}_q)) = \Gamma_q^*(k).$$

PROOF. The inclusion $\Gamma_q^*(k) \subseteq \mathsf{I}(\mathbb{P}^n(\mathbb{F}_q))$ is obvious. For the reverse inclusion, let f be a homogeneous polynomial of degree d such that $f \in \mathsf{I}(\mathbb{P}^n(\mathbb{F}_q))$. By Lemma 3.2, we can write $f = g + \gamma$ for some $g \in \mathfrak{R}_d$ and $\gamma \in \Gamma_q^*(k)_d$. But then $g = f - \gamma$ vanishes on $\mathbb{P}^n(\mathbb{F}_q)$ and so by Lemma 3.1, $g = 0$. Thus $f \in \Gamma_q^*(k)$. □

REMARK 3.4. The result in Theorem 3.3 when $k = \mathbb{F}_q$ has been stated by Bayer-Fluckiger and Serre in [**2**, Lem. 7.2.4], where the proof is left to the reader. As noted earlier, a complete proof was given by Mercier and Rolland [**18**, Thm. 2.1]. Another proof that uses part (i) of Theorem 2.3 appears in the paper of Rentería and Tapia-Recillas [**20**, Prop. 8]. It may be noted that although [**20**] was published in 1997 and [**18**] in 1998, the former was received by the journal in March 1996 and the latter in January 1995. Moreover, [**18**] is included among the references of [**20**].

Unlike in the affine case, it is not true that if $\mathcal{I}$ is a homogeneous ideal of $\mathcal{S}$, then $\mathsf{I}(\mathsf{V}(\mathcal{I})) = \mathcal{I} + \Gamma_q^*(k)$. In fact, while $\mathsf{I}(\mathsf{V}(\mathcal{I}))$ is a radical ideal, the ideal $\mathcal{I} + \Gamma_q^*(k)$ need not be a radical ideal even when $\mathcal{I}$ is a radical ideal of $\mathcal{S}$. To illustrate this, we reproduce the following example from [**3**, Ex. 3.8].

EXAMPLE 3.5. Suppose $n = 1$ and $f(X_0, X_1) := X_0^q X_1 - X_0 X_1^q + X_0^{q+1}$. Consider the principal homogeneous ideal $\mathcal{I} = \langle f(X_0, X_1) \rangle$ of $k[X_0, X_1]$. Note that $\mathcal{I}$ is a radical ideal of $\mathcal{S} = k[X_0, X_1]$ because $\mathcal{S}$ is a UFD and f does not have a multiple root in $\mathbb{P}^1(k)$. Indeed, $f(X_0, X_1) = X_0 g(X_0, X_1)$ where $g(X_0, X_1) := X_0^{q-1} X_1 - X_1^q + X_0^q$ does not have $[0:1]$ as a root and also no multiple root of the form $[1:a]$ since the derivative with respect to X_1 of $g(1, X_1)$ is never zero. On the other hand, $\mathcal{I} + \Gamma_q^*(k) = \mathcal{I} + \langle X_0^q X_1 - X_0 X_1^q \rangle$ contains X_0^{q+1}, but does not contain X_0 (since every nonzero element of $\mathcal{I} + \Gamma_q^*(k)$ has degree $\geq q+1$). Thus $\mathcal{I} + \Gamma_q^*(k)$ is not a radical ideal even though $\mathcal{I}$ is a radical ideal.

REMARK 3.6. The nonavailability of a straightforward analogue of part (ii) of Theorem 2.3 in the projective case makes it harder to find a suitable analogue of the affine $\mathbb{F}_q$-footprint bound (Lemma 2.6). Nonetheless, it is shown in [**3**] how a useful projective $\mathbb{F}_q$-footprint bound can be obtained, and as an application an inequality due to Serre for the number of points of projective hypersurfaces is deduced. This inequality is, in fact, a not-so-straightforward projective analogue of Ore's inequality given in Corollary 2.7; see [**6**] for more on this.

Acknowledgements

The author is grateful to Peter Beelen and Mrinmoy Datta for some helpful conversations related to the topics in this article.

References

[1] E. Arrondo, *Another elementary proof of the Nullstellensatz*, Amer. Math. Monthly **113** (2006), no. 2, 169–171, DOI 10.2307/27641869. MR2203239

[2] E. Bayer-Fluckiger and J.-P. Serre, *Torsions quadratiques et bases normales autoduales* (French), Amer. J. Math. **116** (1994), no. 1, 1–64, DOI 10.2307/2374981. MR1262426

[3] P. Beelen, M. Datta, and S. R. Ghorpade, *Vanishing ideals of projective spaces over finite fields and a projective footprint bound*, Acta Math. Sin. (Engl. Ser.) **35** (2019), no. 1, 47–63, DOI 10.1007/s10114-018-8024-7. MR3917991

[4] C. Carvalho, Applications of results from commutative algebra to the study of certain evaluation codes, Lecture notes of CIMPA Research School on Algebraic Methods in Coding Theory, Sao Paulo, Brazil, July 2017. `https://www.ime.usp.br/~cimpars/notes/sc4_01.pdf`

[5] D. Cox, J. Little, and D. O'Shea, *Ideals, varieties, and algorithms: An introduction to computational algebraic geometry and commutative algebra*, Undergraduate Texts in Mathematics, Springer-Verlag, New York, 1992. MR1189133

[6] M. Datta and S. R. Ghorpade, *On a conjecture of Tsfasman and an inequality of Serre for the number of points of hypersurfaces over finite fields*, Mosc. Math. J. **15** (2015), no. 4, 715–725, DOI 10.17323/1609-4514-2015-15-4-715-725. MR3438829

[7] P. Delsarte, J.-M. Goethals, and F. J. MacWilliams, *On generalized Reed-Muller codes and their relatives*, Information and Control **16** (1970), 403–442. MR0274186

[8] D. Eisenbud, *Commutative algebra: With a view toward algebraic geometry*, Graduate Texts in Mathematics, vol. 150, Springer-Verlag, New York, 1995. MR1322960

[9] K. Goel, D. P. Patil, and J. K. Verma, Nullstellensätze and applications, `arXiv:1809.02818 [math.AC]`, Sept. 2018.

[10] D. Hilbert, *Ueber die Theorie der algebraischen Formen* (German), Math. Ann. **36** (1890), no. 4, 473–534, DOI 10.1007/BF01208503. MR1510634

[11] D. Hilbert, *Ueber die vollen Invariantensysteme* (German), Math. Ann. **42** (1893), no. 3, 313–373, DOI 10.1007/BF01444162. MR1510781

[12] D. Hilbert, *Hilbert's invariant theory papers*, Lie Groups: History, Frontiers and Applications, VIII, Math Sci Press, Brookline, Mass., 1978. Translated from the German by Michael Ackerman; With comments by Robert Hermann. MR512034

[13] J.-R. Joly, *Équations et variétés algébriques sur un corps fini* (French), Enseignement Math. (2) **19** (1973), 1–117. MR0327723

[14] M. Kreuzer and L. Robbiano, *Computational linear and commutative algebra*, Springer, Cham, 2016. MR3559741

[15] D. Laksov, *Radicals and Hilbert Nullstellensatz for not necessarily algebraically closed fields*, Enseign. Math. (2) **33** (1987), no. 3-4, 323–338. MR925995

[16] R. Lidl and H. Niederreiter, *Finite fields*, 2nd ed., Encyclopedia of Mathematics and its Applications, vol. 20, Cambridge University Press, Cambridge, 1997. MR1429394

[17] J. P. May, *Munshi's proof of the Nullstellensatz*, Amer. Math. Monthly **110** (2003), no. 2, 133–140, DOI 10.2307/3647772. MR1952440

[18] D.-J. Mercier and R. Rolland, *Polynômes homogènes qui s'annulent sur l'espace projectif* $\mathrm{P}^m(\mathbf{F}_q)$ (French, with French summary), J. Pure Appl. Algebra **124** (1998), no. 1-3, 227–240, DOI 10.1016/S0022-4049(96)00104-1. MR1600301

[19] R. Munshi, Hilbert's Nullstellensatz, *Bull. Bombay Math. Colloq.* **15** (1999), 20–24.

[20] C. Rentería and H. Tapia-Recillas, *Reed-Muller codes: an ideal theory approach*, Comm. Algebra **25** (1997), no. 2, 401–413, DOI 10.1080/00927879708825862. MR1428786

[21] A. Seidenberg, *Constructions in algebra*, Trans. Amer. Math. Soc. **197** (1974), 273–313, DOI 10.2307/1996938. MR0349648

[22] G. Stengle, *A nullstellensatz and a positivstellensatz in semialgebraic geometry*, Math. Ann. **207** (1974), 87–97, DOI 10.1007/BF01362149. MR0332747

[23] G. Terjanian, *Sur les corps finis* (French), C. R. Acad. Sci. Paris Sér. A-B **262** (1966), A167–A169. MR0194416

[24] O. Zariski, *A new proof of Hilbert's Nullstellensatz*, Bull. Amer. Math. Soc. **53** (1947), 362–368, DOI 10.1090/S0002-9904-1947-08801-7. MR0020075

Department of Mathematics, Indian Institute of Technology Bombay,
Powai, Mumbai 400076, India.
Email address: `srg@math.iitb.ac.in`

Contemporary Mathematics
Volume **738**, 2019
https://doi.org/10.1090/conm/738/14877

Minkowski inequality and equality for multiplicity of ideals of finite length in Noetherian local rings

Kriti Goel, R. V. Gurjar, and J. K. Verma

ABSTRACT. In this exposition of the equality and inequality of Minkowski for multiplicity of ideals, we provide simple algebraic and geometric proofs. Connections with mixed multiplicities of ideals are explained.

1. Introduction

The objective of this expository paper is to present an account of Minkowski's inequality and equality for multiplicity of ideals of finite co-length in Noetherian local rings. These were first investigated by B. Teissier in his Cargèse paper [**28**], in which he proposed conjectures about mixed multiplicities of ideals which imply Minkowski's inequality for multiplicities of ideals. We shall present geometric proofs using an interpretation of multiplicity of an ideal due to C. P. Ramanujam [**22**]. We present a simpler version of the Rees-Sharp proof of Minkowski equality for multiplicities in dimension 2. A proof of Minkowski's equality in dimension ≥ 3 is presented using specialization of the integral closure of ideals due to S. Itoh [**14**] and Hong-Ulrich [**11**]. In his book *"Singular points of hypersurfaces"* [**19**], John Milnor proved the following:

THEOREM 1.1. *Suppose that the origin is an isolated singular point of a complex analytic hypersurface $H = V(f) \subset \mathbb{C}^{n+1}$ and S is an n-dimensional sphere centered at the origin of sufficiently small radius. Define $\varphi(z) : S \setminus H \to S^1$ given by $\varphi(z) = \frac{f(z)}{||f(z)||}$. Then each fiber of φ upto homotopy is a wedge of μ copies of S^n where*

$$\mu = \dim_{\mathbb{C}} \frac{\mathbb{C}\{z_0, z_1, \dots, z_n\}}{(f_{z_0}, f_{z_1}, \dots, f_{z_n})}.$$

The number μ is called the Milnor number of the hypersurface H at the origin. The Milnor number of X at an isolated singular point $x \in X$ will be denoted by $\mu(X, x)$. B. Teissier [**28**] showed that $\mu(X, x)$ is a topological invariant.

2010 *Mathematics Subject Classification.* Primary 13-02, 13B22, 13F30, 13H15, 14C17.
Key words and phrases. multiplicity and mixed multiplicities of ideals, integral closure of ideals, geometric interpretation of multiplicity, $\mathfrak{m}$-valuations.
The first author is supported by a UGC fellowship, Govt. of India.
The third author is the corresponding author.

THEOREM 1.2 (**Teissier, 1973**). *Let (X,x) and (Y,y) be two germs of hyper surfaces with isolated singularity having same topological type. Then*

$$\mu(X,x) = \mu(Y,y).$$

Teissier, in his Cargèse paper [**28**], refined the notion of Milnor number. He replaced it by a sequence of Milnor numbers of intersections of X with general linear subspaces.

THEOREM 1.3 (**Teissier, 1973**). *Let (X,x) be a germ of a hypersurface in $\mathbb{C}^{n+1}$ with an isolated singularity. Let E be an i-dimensional affine subspace of $\mathbb{C}^{n+1}$ passing through x. If E is sufficiently general, then the Milnor number of $X \cap E$ at x is independent of E.*

DEFINITION 1.4. The Milnor number of $X \cap E$, where E is a general linear subspace of dimension i passing through x is called the ith-sectional Milnor number of X. It is denoted by $\mu^{(i)}(X,x)$. These are collected together as

$$\mu^*(X,x) = (\mu^{(n+1)}(X,x), \mu^{(n)}(X,x), \ldots, \mu^{(0)}(X,x)).$$

It is easy to see that $\mu^{(0)}(X,x) = 1$ and $\mu^{(1)}(X,x) = m_x(X) - 1$ where $m_x(X)$ denotes the multiplicity of X at x. Teissier proposed the following:

CONJECTURE 1.5. If the germs of isolated hypersurface singularities (X,x) and (Y,y) have same topological type then

$$\mu^*(X,x) = \mu^*(Y,y).$$

This conjecture contains the following conjecture of O. Zariski [**30**].

CONJECTURE 1.6 (**Zariski, 1971**). For topologically equivalent isolated singularities of hypersurfaces, $m_x(X) = m_y(Y)$.

Zariski conjecture has been established by Lê Dũng Tráng for plane curves in [**29**]. Teissier's Conjecture 1.5 was disproved in 1975 by J. Briançon and J.-P. Speder [**5**]. Another conjecture made by Teissier was about the log-convexity of the sectional Milnor numbers:

CONJECTURE 1.7 (**Teissier, 1973**). Is it true that

$$\frac{\mu^{(n+1)}}{\mu^{(n)}} \geq \frac{\mu^{(n)}}{\mu^{(n-1)}} \geq \cdots \geq \frac{\mu^{(1)}}{\mu^{(0)}}.$$

Since all the sectional Milnor numbers are positive integers, these inequalities are equivalent to asking if for all $i = 1, 2, \ldots, n$,

$$\log \mu^{(i)} \leq \frac{1}{2}\left(\log \mu^{(i-1)} + \log \mu^{(i+1)}\right).$$

In other words the sequence $\mu^{(0)}, \mu^{(1)}, \ldots, \mu^{(n+1)}$ is a log-convex sequence. In an appendix to the paper by David Eisenbud and Harold I. Levine [**9**], Teissier answered this in the affirmative in a much more general setting. Following a suggestion of Hironaka, he considered the Bhattacharya function [**3**] of two ideals to identify the sectional Milnor numbers with mixed multiplicities of ideals. First we recall these notions and related results.

THEOREM 1.8. *Let $(R, \mathfrak{m})$ be a Noetherian local ring of dimension d and let I be an $\mathfrak{m}$-primary ideal. The Hilbert function of I, $H_I(n) := \ell(R/I^n)$ is given by the Hilbert polynomial of I,*

$$P_I(x) = e_0(I)\binom{x+d-1}{d} - e_1(I)\binom{x+d-2}{d-1} + \cdots + (-1)^d e_d(I)$$

for all large n. The coefficients $e_0(I), e_1(I), \ldots, e_d(I)$ are integers. The leading coefficient $e_0(I)$ is a called the multiplicity of I and is denoted by $e(I)$.

The above result of P. Samuel was extended for two $\mathfrak{m}$-primary ideals by Phani Bhushan Bhattacharya in 1957 [**3**].

THEOREM 1.9 (**P. B. Bhattacharya**). *Let I and J be $\mathfrak{m}$-primary ideals of a d-dimensional Noetherian local ring $(R, \mathfrak{m})$. Then the function*

$$H(r,s) = \ell\left(\frac{R}{I^r J^s}\right)$$

is given by a polynomial $P(r,s)$ of degree d, when r and s are large.

The polynomial $P(r,s)$, called the *Bhattacharya polynomial* of I and J, can be written as

$$P(r,s) = \frac{1}{d!}\sum_{i=0}^{d}\binom{d}{i} e_i(I|J) r^{d-i} s^i + \text{ terms of degree } < d$$

where $e_0(I|J), e_1(I|J), \ldots, e_d(I|J)$ are called the *mixed multiplicities* of I and J. David Rees showed that $e_0(I|J) = e(I)$ and $e_d(I|J) = e(J)$. The other mixed multiplicities are also multiplicities of certain system of parameters [**28**].

THEOREM 1.10 (**J. J. Risler and Teissier**). *Each mixed multiplicity $e_i(I|J)$ is the multiplicity of an ideal generated by $(d-i)$ general elements from I and i general elements from J.*

Risler and Teissier developed the theory of mixed multiplicities for a finite family of $\mathfrak{m}$-primary ideals in a local ring $(R, \mathfrak{m})$. Since we need this theory for only two ideals, we shall consider it only in this case for the sake of simplicity.

Rees introduced joint reductions [**24**] in order to find systems of parameters whose multiplicities are the mixed multiplicities. Recall that an ideal $J \subset I$ is called a reduction of I if there is an n such that $JI^n = I^{n+1}$. Let $I_1, I_2, \ldots, I_d$ be $\mathfrak{m}$-primary ideals of R. A sequence of elements $x_1 \in I_1, \ldots, x_d \in I_d$ is called a joint reduction of the sequence of ideals $(I_1, \ldots, I_d)$ if $\sum_{i=1}^d x_i I_1 I_2 \cdots I_{i-1} I_{i+1} \cdots I_d$ is a reduction of $I_1 I_2 \cdots I_d$. Let $(I^{[m]}, J^{[n]})$ denote the sequence of ideals in which the first m ideals are I and the next n ideals are J.

THEOREM 1.11 (**Rees, 1984**). *Let $x_1, x_2, \ldots, x_d$ be a joint reduction of the sequence of ideals $(I^{[d-i]}, J^{[i]})$ then $e_i(I|J) = e(x_1, x_2, \ldots, x_d)$.*

THEOREM 1.12 (**Teissier, 1973**). *Let $X = V(f)$ be an analytic hypersurface in $\mathbb{C}^{n+1}$ with an isolated singularity at the origin. Then for all $i = 0, 1, \ldots, n+1$,*

$$\mu^{(i)}(X, 0) = e_i(\mathfrak{m}|J(f)).$$

This result led Teissier to propose the following conjecture, which if true, implies that the sequence $\mu^*(X, x)$ is log-convex.

CONJECTURE 1.13 (**Teissier's Second Conjecture**). Let I and J be $\mathfrak{m}$-primary ideals of a Noetherian local ring of dimension $d \geq 2$. Put $e_i = e_i(I|J)$. Then

$$\frac{e_1}{e_0} \leq \frac{e_2}{e_1} \leq \frac{e_3}{e_2} \leq \cdots \leq \frac{e_d}{e_{d-1}}.$$

Using the Bhattacharya polynomial, we can see that for all $r, s \in \mathbb{N}$,

$$e(I^r J^s) = e(I)r^d + \binom{d}{1} e_1(I|J)r^{d-1}s + \cdots + e_i(I|J)\binom{d}{i} r^{d-i}s^i + \cdots + e(J)s^d.$$

$$e(IJ) = e(I) + \binom{d}{1} e_1(I|J) + \cdots + e_i(I|J)\binom{d}{i} + \cdots + e(J).$$

Teissier compared the expansion

$$(e(I)^{1/d} + e(J)^{1/d})^d = e(I) + \cdots + \binom{d}{i} e(I)^{(d-i)/d} e(J)^{i/d} + \cdots + e(J)$$

with the formula for $e(IJ)$ and proposed his first conjecture:

CONJECTURE 1.14 (**Teissier's First Conjecture**). Let $(R, \mathfrak{m})$ be a Noetherian local ring of dimension d and I, J be $\mathfrak{m}$-primary ideals. Then for all $i = 0, 1, \ldots, d$,

$$e_i(I|J)^d \leq e(I)^{d-i} e(J)^i.$$

Teissier asked if the Minkowski type inequality is true for the multiplicity of ideals:

$$e(IJ)^{1/d} \leq e(I)^{1/d} + e(J)^{1/d}.$$

Henceforth, we call this Minkowski's inequality. Note that the second conjecture implies the first and the first conjecture implies that Minkowski's inequality is true for multiplicities of ideals. He proved Minkowski's inequality for multiplicities in [**27**] for Cohen-Macaulay local algebras over an algebraically closed field of characteristic zero in dimension two and then used superficial elements to prove it in any dimension. D. Rees and R. Y. Sharp [**26**] proved it in all Noetherian local rings.

THEOREM 1.15 (**Rees-Sharp, 1978**). *The first and the second conjecture of Teissier are true for all Noetherian local rings. In particular, Minkowski's inequality for multiplicity is true in all Noetherian local rings.*

It is natural to ask under what conditions Minkowski inequality is an equality. This requires the concept of integral closure of an ideal, which we recall. Let I be an ideal of a commutative ring R. We say that $x \in R$ is *integral over* I if there exist $a_i \in I^i$ for $i = 1, \ldots, n$ such that $x^n + a_1 x^{n-1} + \cdots + a_n = 0$. The integral closure of I is the ideal

$$\overline{I} = \{x \in R \mid x \text{ is integral over } I\}.$$

Recall that a Noetherian local ring $(R, \mathfrak{m})$ is called *quasi-unmixed* if for all minimal primes $\mathfrak{p}$ of the $\mathfrak{m}$-adic completion $\hat{R}$, we have $\dim \hat{R}/\mathfrak{p} = \dim R$.

THEOREM 1.16. *Let $(R, \mathfrak{m})$ be quasi-unmixed and $d \geq 2$. Then*

$$e(IJ)^{1/d} = e(I)^{1/d} + e(J)^{1/d} \iff \overline{I^r} = \overline{J^s} \text{ for some } r, s \in \mathbb{N}$$

$$\iff \frac{e_i}{e_{i-1}} = \frac{r}{s} \text{ for all } i = 1, \ldots, d.$$

It was proved by Teissier for complex analytic Cohen-Macaulay algebras in arbitrary dimension, by reduction to the case of dimension 2, which appeared in the 1978 Conference Proceedings dedicated to C. P. Ramanujam [**27**]. Rees and Sharp proved it for $d = 2$ and D. Katz reduced the proof for dimension ≥ 3 to dimension 2 in [**15**].

June Huh [**12**] settled a long standing conjecture in graph theory about chromatic polynomials using sectional Milnor numbers and mixed multiplicities. Some of the results he proved are: (1) If J is an ideal of a standard graded domain over an algebraically closed field generated by elements of the same degree, then the mixed multiplicities of $\mathfrak{m}$ and J form a log-concave sequence of nonnegative integers with no internal zeros. (2) Let $h \in \mathbb{C}[x_0, x_1, \ldots, x_n]$ be a homogeneous polynomial of positive degree. Put

$$V(h) = \{p \in \mathbb{P}^n \mid h(p) = 0\} \text{ and } D(h) = \mathbb{P}^n \setminus V(h).$$

Let $\mu^i(h)$ be the ith mixed multiplicity of $\mathfrak{m}$ and $J(h)$. Then the Euler Characteristic of $D(h)$ is given by

$$\chi(D(h)) = \mu^0(h) - \mu^1(h) + \cdots + (-1)^n \mu^n(h).$$

(3) The numbers $\mu^i(h)$ form a log-concave sequence of nonnegative integers with no internal zeros for any h. He uses above results and various properties of matroids, mixed multiplicities, mixed volumes of convex bodies and Milnor numbers to show that the coefficients of the chromatic polynomial of a graph form a log-concave sequence.

Recently, Minkowski's inequality has been proved for non-Noetherian filtrations of ideals by Dale Cutkosky, Parangama Sarkar and Hema Srinivasan [**7**]. Multiplicity of non-Noetherian filtrations of ideals has been investigated by many authors. The most general result for such filtrations was proved by Cutkosky [**6**]. Let $N(\hat{R})$ denote the nilradical of $\hat{R}$.

THEOREM 1.17 (**Cutkosky, 2004**). *Let $\mathcal{F} = \{I_n\}$ be a filtration of $\mathfrak{m}$-primary ideals of a d-dimensional Noetherian local ring R. Then the limit*

$$\lim_{n\to\infty} \frac{\ell(R/I_n)}{n^d} \textit{ exists } \iff \dim N(\hat{R}) < d.$$

This limit has been investigated by several authors, for example, by Ein, Lazarsfeld and Smith [**8**] and Mustaţă [**20**].

THEOREM 1.18 (**Cutkosky-Sarkar-Srinivasan, 2018**). *Let $(R, \mathfrak{m})$ be a d-dimensional Noetherian local ring and let $\{I(1)_n\}, \ldots, \{I(r)_n\}$ be filtrations of $\mathfrak{m}$-primary ideals. Let M be a finitely generated R-module. If $\dim N(\hat{R}) < d$, then the function*

$$P(n_1, n_2, \ldots, n_r) = \lim_{m\to\infty} \frac{\ell(M/I(1)_{mn_1} \cdots I(r)_{mn_r} M)}{m^d}$$

is a real homogeneous polynomial $G(n_1, \ldots, n_r)$ of degree d for all $n_1, \ldots, n_r \in \mathbb{N}$ and it can be written as

$$G(n_1, n_2, \ldots, n_r) = \sum_{d_1+d_2+\cdots+d_r=d} e_R(I(1)^{[d_1]}, \ldots, I(r)^{[d_r]}; M) \frac{n_1^{d_1} \cdots n_r^{d_r}}{d_1! \cdots d_r!}.$$

The integer $e_R(I(1)^{[d_1]}, \ldots, I(r)^{[d_r]}; M)$ is called the mixed multiplicity of the $\mathbb{N}^r$-graded filtration $\{I(1)_{n_1} \cdots I(r)_{n_r} \mid n_1, \ldots, n_r \in \mathbb{N}\}$ of the type $(d_1, \ldots, d_r)$. Many of the classical properties of mixed multiplicities of ideals continue to be true in this setting. In particular, Minkowski's inequalities are proved in [**7**].

Several historical facts mentioned in this paper are taken from the presentation of Patrick Popescu-Pampu made in the conference *Singular Landscapes,* held in honour of Teissier's 70th birthday in 2015 [**21**].

2. C.P. Ramanujam's result, Teissier's Conjecture and a related inequality

In this section, we use basic intersection theory of divisors on smooth algebraic varieties to give short proofs of three results: (1) C.P. Ramanujam's geometric interpretation of multiplicity, (2) Minkowski's inequality for multiplicity, and (3) necessary and sufficient conditions for Minkowski's equality. B. Teissier gave geometric proofs of (2) and (3) in [**27**]. We believe that the proofs presented here are more accessible to a young reader.

For the sake of exposition, we will assume that k is an algebraically closed field. We will only consider local rings whose residue field is isomorphic to k. By an algebraic local ring we mean either a local ring of an algebraic variety over k at a maximal ideal, or a local analytic ring of the form $\mathbb{C}\{Z_1, Z_2, \ldots, Z_n\}/P$, where $\mathbb{C}\{Z_1, \ldots, Z_n\}$ is the convergent power series ring over the complex field and P is a prime ideal. We assume that the reader is familiar with the basic notions in algebraic geometry.

Intersection Theory. Our proofs of the results depend crucially on the intersection theory of curves on a smooth surface, or divisors in higher dimensional smooth varieties. We will recall this basic theory, mostly without proofs. An excellent source for the detailed properties of intersection theory in arbitrary dimension is the book [**10**].

Let X be a smooth irreducible surface over k. Let C, D be (possibly non-reduced and reducible) curves on X without a common irreducible component. For a point $p \in C \cap D$, there are functions r, s in the local ring R of X at p such that C, D are scheme-theoretically defined by r, s respectively. Then $\dim_k R/(r, s)$ is called the *intersection multiplicity* of C and D at p, denoted by $i(C, D; p)$. It can be proved that if C_m, D_n are all the irreducible components of C, D respectively passing through p (counting multiplicities), then $i(C, D; p) = \Sigma_{m,n}\, i(C_m, D_n; p)$. A useful result in this connection is the following lemma. Recall that the residue field k of local rings occuring in the result below is assumed to be algebraically closed.

LEMMA 2.1 (Abhyankar, [**1**]). *Let (R, M) be a Cohen Macaulay 1-dimensional algebraic or local analytic ring, and let $\overline{R}$ be the normalization of R in its total quotient ring. Then for any nonzerodivisor $r \in R$, $\dim_k R/rR = \dim_k \overline{R}/r\overline{R}$.*

PROOF. Let $\overline{R}/r\overline{R} \supset I_1 \supset \ldots \supset I_l = (0)$ be a Jordan-Hölder sequence of ideals. Then each I_j/I_{j+1} is a 1-dimensional k-vector space and hence an R/M-module of length 1. Therefore, this is also a Jordan-Hölder sequence of R-modules. This implies

$$l_R(\overline{R}/r\overline{R}) = l_{\overline{R}}(\overline{R}/r\overline{R}).$$

Consider $rR \subset R \subset \overline{R}$ and $rR \subset r\overline{R} \subset \overline{R}$. Then

$$\begin{aligned} l_R(\overline{R}/R) + l_R(R/rR) = l_R(\overline{R}/rR) &= l_R(\overline{R}/r\overline{R}) + l_R(r\overline{R}/rR) \\ &= l_R(\overline{R}/r\overline{R}) + l_R(\overline{R}/R). \end{aligned}$$

Note that $l_R(r\overline{R}/rR) = l_R(\overline{R}/R)$ as r is a nonzerodivisor on R. This shows that $l_R(R/rR) = l_R(\overline{R}/r\overline{R}) = l_{\overline{R}}(\overline{R}/r\overline{R})$. □

Let X, C, D be as above. Then $C \cap D$ is a finite set of points $p_1, \ldots, p_m$. We define

$$C.D := \Sigma_j \ i(C, D; p_j).$$

Now let $C := \Sigma_m a_m C_m$, $D := \Sigma_n b_n D_n$ be formal sums of irreducible curves C_m, D_n on X such that no C_m is equal to any D_n and a_m, b_n are arbitrary non-zero integers. Such a formal sum is called a *divisor* on X. We define

$$C.D := \Sigma_{m,n} \ a_m.b_n C_m.D_n.$$

Let X be a smooth projective surface and C a reduced, irreducible curve on X. Our aim is to define $C.C$. We can find a non-zero rational function f on X such that $(f) + C$ is a divisor D on X such that the supports of C, D have no common curve. We define $C.C := C.D$. Using the well-known result that the orders of zeros and poles of a non-zero rational function on a smooth projective curve are equal we can show that $C.D$ is well-defined.

Let X be a smooth projective surface and C, D divisors on X. We say that C, D are *linearly equivalent* (or *rationally equivalent*), written $C \sim D$, if the divisor $C - D = (r)$ for some non-zero rational function r on X. Now if $C_1 \sim C_2$ then for any divisor D on X, we have $C_1.D = C_2.D$.

If $\pi : Y \to X$ is a proper surjective morphism between smooth projective surfaces, then for any divisors C, D on X we have $\pi^*C.\pi^*D = degree\ \pi.(C.D)$. Here, π^*C, π^*D are the scheme-theoretic pull-backs of C, D by π.

Let $\pi : Y \to X$ be a surjective morphism between normal projective surfaces. Let $E_1, \ldots, E_m$ be all the irreducible curves on Y such that $\pi(E_j)$ is a point in X for every j. We call E_j the exceptional curves for the morphism π. It can be shown that for any divisor C on X, we have $\pi^*C.E_j = 0$ for each j.

Now we come to an important basic result in the intersection theory on surfaces due to Patrick Du Val.

LEMMA 2.2. *Let X be a normal projective surface, Y a smooth projective surface and $f : Y \to X$ be a surjective morphism. Let $E_1, \ldots, E_m$ be all the exceptional curves on Y for f. Then the intersection form on $\cup E_j$ is negative definite.*

PROOF. For simplicity, we will assume that all the curves E_j map to the same point $p \in X$. The general case is similar. Let r be a regular function on X in a neighborhood of p. Then on Y we have $D := (r)_Y = C + \Sigma m_j E_j$, where C is the part of the divisor $(r)_Y$ which does not contain any E_j in its support. Then $(r)_Y.E_j = 0$ for every j. Hence

$$D.E_j = C.E_j + E_j.(\Sigma m_l E_l)$$

for every $j > 0$. Since r is regular at p, we see that $C.E_j \geq 0$ for every $j > 0$. Also, it can be shown easily that $C.E_j > 0$ for some $j > 0$. Thus, $(\Sigma m_j E_j).E_l \leq 0$ for every l and strict inequality holds for some l. We will show that this implies that the intersection form on $\cup E_j$ is negative definite.

Consider the symmetric quadratic form on an m-dimensional real vector space, with basis $x_1, \ldots, x_m$ given by $\Sigma\ \alpha_{ij} m_i m_j x_i.x_j$, where $\alpha_{ij} = E_i.E_j$. It suffices to show that this form is negative definite. We have
(1) $\alpha_{ij} \geq 0$ if $i \neq j$,
(2) $\Sigma_i \alpha_{ij} = (\Sigma m_i C_i).m_j E_j \leq 0$ for every j, and
(3) $\Sigma_i \alpha_{ij} < 0$ for some j.
From this, we get

$$\Sigma_{ij}\ \alpha_{ij} x_i x_j = \Sigma_j (\Sigma_i\ \alpha_{ij}) x_j^2 - \Sigma_{i<j}\ \alpha_{ij}(x_i - x_j)^2.$$

This shows the negative semi-definiteness of the intersection form. If the R.H.S. is 0 for some real values $x_1, \ldots, x_m$ then (3) shows that $x_j = 0$ if $\Sigma_i \alpha_{ij} < 0$. But then $x_i = x_j$ for all i, j and hence the result follows. □

The inequalities. Let (R, M) be an algebraic or complex analytic local domain of dimension d with $R/M \cong k$. In the complex analytic case, $k = \mathbb{C}$. Let $I \subset R$ be an M-primary ideal, and let $e(I)$ be the multiplicity of I. Let $\pi : X \to \operatorname{Spec} R$ be a resolution of singularities such that $\pi^* I$ is an invertible sheaf of ideals on X. If $\operatorname{char} k = 0$, or $d = 2$, or $d = 3$ and $\operatorname{char} k > 5$, then such a resolution of singularities exists [**16**],[**2**] respectively. Then $\pi^* I$ defines an effective divisor $D = \Sigma n_i D_i$ on X, where D_i are irreducible divisors on X. The next result is the geometric interpretation of multiplicity proved by C.P. Ramanujam [**22**].

THEOREM 2.3. *$e(I) = (-1)^{d-1}.D^n$, where $D^n = D.D \ldots D$ n times.*

PROOF. We can find a minimal reduction $(x_1, \ldots, x_d)$ of I. By definition, $(x_1, \ldots, x_d)I^l = I^{l+1}$ for all l large and $e((x_1, \ldots, x_d)) = e(I)$. It is a standard result that $k[[x_1, \ldots, x_d]] \subset \hat{R}$ is a Noether normalization of degree $e(I)$, where $\hat{R}$ is the completion of R with respect to M. This gives a finite morphism

$$\operatorname{Spec} \hat{R} \to S := \operatorname{Spec}(k[[x_1, \ldots, x_d]])$$

of degree $e(I)$.

Let $\sigma : \tilde{S} \to S$ be the monoidal transform with center $(x_1, \ldots, x_d)$. Then the scheme-theoretic inverse image of the closed point of $\operatorname{Spec} S$ is a reduced divisor $E \cong \mathbf{P}^{d-1}$, and $E.E \ldots E = (-1)^{d-1}$.

We can assume (by further blowing ups on X, if necessary) that there is a proper morphism of degree $e(I)$, say $\tilde{\pi} : X \to \tilde{S}$. By the property of minimal reduction, $(x_1, \ldots, x_d)$ and I generate the same ideal sheaf on X. Thus, $\tilde{\pi} E = D$. By projection formula, $(\tilde{\pi}^* E)^d = (-1)^{d-1} e(I)$, i.e., $D^n = (-1)^{d-1} e(I)$. □

REMARK 2.4. (1) Let (R, M) be the local ring of a 2-dimensional rational singular point. Let $\pi : X \to \operatorname{Spec} R$ be a resolution of singularities. It is known

that π^*M is an invertible sheaf of ideals. M. Artin proved that the divisor of zeros of this ideal sheaf is the fundamental cycle Z for π. By Theorem 2.3, $Z^2 = -e(R)$. This is one of the results proved by Artin about rational surface singular points. Hence C.P. Ramanujam's geometric interpretation of multiplicity is a vast generalization of Artin's result about multiplicity of a rational surface singularity.
(2) Let (R, M) be a normal algebraic local domain and I be an M-primary ideal in R. Let $f : X \to \operatorname{Spec} R$ be a resolution of singularities such that f^*I, $f^*\overline{I}$ are locally principal, where $\overline{I}$ is the integral closure of I in R. Then $f^*I = f^*\overline{I}$. Therefore, if $I \subset J$ are M-primary ideals such that f^*I, $f^*\overline{I}$, f^*J and $f^*\overline{J}$ are locally principal and f^*I, f^*J define the divisors D and E in X. Clearly, $\overline{I} \subset \overline{J}$. It will be proved in the course of the proof of Theorem 9.4 that $D = E$ implies that $\overline{I} = \overline{J}$.
(3) Assume that $d = 2$. Then the intersection form of $\cup D_i$ is negative definite. In particular, the irreducible components $D_1, D_2, \ldots$ are rationally independent. If $k = \mathbb{C}$, then the Chern classes $[D_1], [D_2], \ldots$ are even $\mathbb{Q}$-independent in the rational homology of X. This will be used in the proof of the next result.

THEOREM 2.5. *Let (R, M) be as above and $d = 2$. Let I, J be M-primary ideals. Then we have*
(1) $e(IJ)^{1/2} \leq e(I)^{1/2} + e(J)^{1/2}$.
(2) *Equality holds if and only if the integral closures $\overline{I^r} = \overline{J^s}$ for some $r, s \geq 1$.*

PROOF. Let $\pi : X \to \operatorname{Spec} R$ be a resolution of singularities such that π^*I, π^*J are locally invertible sheaves of ideals. By Theorem 2.3, if D, E are the divisors on X defined by π^*I, π^*J then

$$D^2 = -e(I), E^2 = -e(J), (D+E)^2 = -e(I.J).$$

We want to show that

$$(-(D+E)^2)^{1/2} \leq (-D^2)^{1/2} + (-E^2)^{1/2}.$$

By squaring, this is equivalent to

$$-(D+E)^2 \leq -D^2 - E^2 + 2(-D^2)^{1/2}(-E^2)^{1/2},$$

i.e., $D.E \leq (-D^2)^{1/2}(-E^2)^{1/2}$, or $(D.E)^2 \leq (-D^2).(-E^2) = D^2.E^2$. Since the intersection matrix of D, E is negative definite, the result follows.
Now suppose that the equality holds. Then D, E are rationally dependent divisors. Hence there are positive integers r, s such that $rD = sE$. This again uses the negative definiteness of the intersection form. Since the integral closures $\overline{I^r}, \overline{J^s}$ are the unique largest ideals which define rD, sE respectively, we get $\overline{I^r} = \overline{J^s}$. □

THEOREM 2.6. *Let (R, M) be as in Theorem 2.5 with $d = 2$. Let I, J be M-primary ideals. Then we have*
(1) $e(IJ) \leq 2e(I) + 2e(J)$.
(2) *Equality holds if and only if $\overline{I} = \overline{J}$.*

PROOF. We want to prove

$$-(D+E)^2 \leq 2(-D^2) + 2(-E^2).$$

This is equivalent to $D^2 + E^2 - 2D.E \leq 0$, i.e. $(D-E)^2 \leq 0$. By the negative definiteness of the intersection form this follows. If equality holds, then $D = E$, i.e. $\overline{I} = \overline{J}$. In this case, $e(I^2) = 4e(I)$. Since $2D = \pi^*(I^2)$, this follows from Theorem 2.3. □

3. Bhattacharya function of two ideals

Let $(R, \mathfrak{m})$ be a d-dimensional Noetherian local ring and I, J be $\mathfrak{m}$-primary ideals. Recall that $H(r,s) = \ell(R/I^r J^s)$ is called the Bhattacharya function of I and J. We prove

THEOREM 3.1. *The function $H(r,s)$ is given by a polynomial of degree d for all large $r, s \in \mathbb{N}$. If we write this polynomial as*

$$P(r,s) = \frac{1}{d!}\left\{e_0(I|J)r^d + \cdots + \binom{d}{i} e_i(I|J) r^{d-i}s^i + \cdots + e_d(I|J)s^d\right\} + \cdots$$

then $e_i(I|J)$, for all $i = 0, \ldots, d$, are positive integers.

In order to prove the theorem, we follow the arguments as given in [**28**, Proposition 2.1] by Risler-Teissier. For this, we need to first define superficial element.

DEFINITION 3.2. Let R be a Noetherian ring and I, J be ideals in R. An element $x \in I$ is superficial for I, J if there exists $c > 0$ such that for $r \geq c$ and $s \geq 0$,

$$(I^r J^s : x) \cap I^c J^s = I^{r-1} J^s.$$

THEOREM 3.3. (Existence of superficial elements) *If $(R, \mathfrak{m})$ is a Noetherian local ring with infinite residue field, then superficial element for I, J exist. Explicitly, there exists a non-empty Zariski-open subset U of $I/\mathfrak{m}I$ and $c \in \mathbb{N}$ such that for any $x \in I$ with image in U, for all $r \geq c$ and all $s \geq 0$,*

$$(I^r J^s : x) \cap I^c J^s = I^{r-1} J^s.$$

If I is not contained in the prime ideals $\mathfrak{p}_1, \ldots, \mathfrak{p}_t$ of R, x can be chosen to avoid the same prime ideals.

For geometric interpretation of superficial element in the case of one ideal, we refer the reader to a paper by Romain Bondil [**4**].

One often studies Hilbert polynomials in low dimension and then uses induction on $\dim R$. Superficial elements allow us to pass to lower dimensions. This is possible due to the next result.

THEOREM 3.4. *Let $x \in I$ be superficial for the pair (I, J) with respect to the set of minimal primes of R. Then for all large r and s,*

(1) $\dfrac{I^r J^s : x}{I^{r-1} J^s} \simeq (0 : x)$

(2) $\ell\left(\dfrac{R}{I^r J^s}\right) - \ell\left(\dfrac{R}{I^{r-1} J^s}\right) = \ell\left(\dfrac{R}{I^r J^s + (x)}\right) - \ell((0 : x)).$

PROOF. By Artin-Rees lemma, there exists $(m, n) \in \mathbb{N}^2$ such that for all $(r, s) \geq (m, n)$,

$$I^r J^s \cap (x) \subseteq x I^{r-m} J^{s-n}.$$

Let $a \in R$ such that $ax \in I^r J^s \cap (x)$. Then $ax \in x I^{r-m} J^{s-n}$. Write $ax = xp$ for some $p \in I^{r-m} J^{s-n}$. Then $x(a - p) = 0$ and hence $a \in I^{r-m} J^{s-n} + (0 : x)$. Thus

$$I^r J^s : x \subseteq I^{r-m} J^{s-n} + (0 : x).$$

Since $x \in I$ is superficial with respect to (I, J) there is a $c > 0$ such that for all $r \geq c$ and all $s \geq 0$, $(I^r J^s : x) \cap I^c J^s = I^{r-1} J^s$. As I is $\mathfrak{m}$-primary, there exists

$k > 0$ such that $I^k \subseteq I^c J^n$ and hence $I^k I^{r-m} J^{s-n} \subseteq I^c J^s$ for all $(r,s) \geq (m,n)$. It follows that for all $r \geq m+k$, $s \geq n$,

$$(I^r J^s : x) \subseteq (0 : x) + I^c J^s.$$

Hence for all $r \geq m+k$, $s \geq n$,

$$(I^r J^s : x) = (I^r J^s : x) \cap ((0 : x) + I^c J^s) = (0 : x) + I^{r-1} J^s.$$

Therefore for large r and s,

$$\frac{I^r J^s : (x)}{I^{r-1} J^s} = \frac{(0 : x) + I^{r-1} J^s}{I^{r-1} J^s} \simeq \frac{(0 : x)}{(0 : x) \cap I^{r-1} J^s}.$$

Consider

$$\begin{aligned} (0 : x) \cap I^{r-1} J^s \subseteq \left(\bigcap_{r \geq 1} (I^r J^s : x) \right) \cap I^{r-1} J^s &\subseteq \left(\bigcap_{r \geq 1} (I^r J^s : x) \right) \cap I^c J^s \\ &= \bigcap_{r \geq 1} I^{r-1} J^s \\ &\subseteq \bigcap_{r \geq 1} I^{r-1} = (0). \end{aligned}$$

Hence it follows that for large r and s, $(I^r J^s : (x))/I^{r-1} J^s \simeq (0 : x)$. By avoiding the minimal associated primes of R, we ensure $\dim R/(x) = \dim R - 1$. For the second assertion, use the exact sequence of R-modules,

$$0 \longrightarrow \frac{I^r J^s : (x)}{I^{r-1} J^s} \longrightarrow \frac{R}{I^{r-1} J^s} \xrightarrow{\mu_x} \frac{R}{I^r J^s} \longrightarrow \frac{R}{I^r J^s + (x)} \longrightarrow 0.$$

□

Proof of Theorem 3.1: Proceed by induction on $\dim R = d$. If $d = 0$, then for large r and s, $I^r J^s = 0$. Hence $H(r,s) = \ell(R) = P(r,s)$ for large r, s. Thus, $P(r,s)$ is a degree zero rational polynomial. Let $d > 1$ and $\{\mathfrak{p}_1, \ldots, \mathfrak{p}_t\}$ be the set of minimal primes of R. As I is $\mathfrak{m}$-primary, it follows that $I \not\subseteq \cup_{i=1}^t \mathfrak{p}_i$. Using Theorem 3.3, there exists $x \in I \setminus \cup_{i=1}^t \mathfrak{p}_i$ and $c > 0$ such that for all $r \geq c$ and $s \geq 0$,

$$(I^r J^s : x) \cap I^c J^s = I^{r-1} J^s.$$

From Theorem 3.4,

$$H(R/(x), (r,s)) = H(R, (r,s)) - H(R, (r-1, s)) + \ell(0 : x).$$

By the choice of x, $\dim R/(x) = d-1$ and by induction hypothesis, $H(R/(x), (r,s))$ is a rational polynomial $Q(R/(x), (r,s))$ of degree $d-1$ for large r, s such that every monomial of total degree $d-1$ in the polynomial has a positive coefficient. Let $H(R/(x), (r,s)) = Q(R/(x), (r,s))$ for all $r \geq r_0$, $s \geq s_0$. One can now conclude that for $r \geq r_0, s \geq s_0$,

$$H(R, (r,s)) = H(R, (r_0, s)) + \sum_{i=r_0+1}^{r} \left(Q\left(\frac{R}{(x)}, (i,s) \right) - \ell(0 : x) \right).$$

Since $H(R, (r_0, s))$ and $\sum_{i=r_0+1}^{r} Q(R/(x), (i,s))$ are rational polynomials of degree d (see [**13**, Lemma 11.1.2]) with every monomial of total degree d having a positive coefficient, we are done.

REMARK 3.5. Let $e'_0, \ldots, e'_d$ denote the mixed multiplicities of I and J as ideals in $R/(x)$. Using part (2) of Theorem 3.4, it follows that, $e_0 = e'_0, \ldots, e_{d-1} = e'_{d-1}$.

Given ideals I, J of a ring R, Rees introduced the notion of a Rees superficial element for the pair (I, J) in [**24**].

DEFINITION 3.6. Let R be a Noetherian ring and I, J be ideals in R. An element $x \in I$ is Rees superficial for I, J if there exists $c > 0$ such that for $r \geq c$ and $s \geq 0$,

$$I^r J^s \cap (x)R = xI^{r-1}J^s.$$

LEMMA 3.7 ([**24**, Lemma 1.2]). (Existence of Rees superficial element) *Let $(R, \mathfrak{m})$ be a local ring with $R/\mathfrak{m}$ infinite Let I, J be ideals of R and let $\mathfrak{p}_1, \ldots, \mathfrak{p}_t$ be prime ideals which do not contain IJ. Then there exist $x \in I \setminus (\mathfrak{p}_1 \cup \cdots \cup \mathfrak{p}_t \cup \mathfrak{m}I)$ and $c > 0$ such that for all $r \geq c$ and $s \geq 0$,*

$$I^r J^s \cap (x)R = xI^{r-1}J^s.$$

THEOREM 3.8. *Let I, J be $\mathfrak{m}$-primary ideals of a ring $(R, \mathfrak{m})$ and $x \in I$ be a Rees superficial element for the pair (I, J). If $\operatorname{ann}(x) \subseteq \operatorname{ann}(I)$, then for all large r and s,*

$$\ell\left(\frac{R}{I^r J^s}\right) - \ell\left(\frac{R}{I^{r-1}J^s}\right) = \ell\left(\frac{R}{I^r J^s + (x)}\right) - \ell((0:x)).$$

PROOF. As x is Rees superficial for (I, J), there exists $c > 0$ such that for all $r \geq c$ and $s \geq 0$, $I^r J^s \cap (x) = xI^{r-1}J^s$. Using the arguments as in proof of Theorem 3.4, it is sufficient to show that for large r, s,

$$\frac{I^r J^s : (x)}{I^{r-1}J^s} \simeq (0:x).$$

Indeed, let $b \in I^r J^s : (x)$. Then for all $r \geq c$ and $s \geq 0$, $bx \in I^r J^s \cap (x) = xI^{r-1}J^s$. Write $bx = xu$, for some $u \in I^{r-1}J^s$. This implies that $(b - u) \in (0:x)$ and hence for all $r \geq c$ and $s \geq 0$, $b \in (0:x) + I^{r-1}J^s$. Therefore for all $r \geq c$ and $s \geq 0$,

$$\frac{I^r J^s : (x)}{I^{r-1}J^s} = \frac{(0:x) + I^{r-1}J^s}{I^{r-1}J^s} \simeq \frac{(0:x)}{(0:x) \cap I^{r-1}J^s}.$$

Using Artin-Rees lemma, there exist r_0, s_0 such that for all $r \geq r_0$ and $s \geq s_0$,

$$(0:x) \cap I^{r-1}J^s \subseteq (0:x)I^{r-1-r_0}J^{s-s_0} = 0.$$

The last equality follows as $(0:x) \subseteq (0:I)$. Hence it follows that for large r and s, $(I^r J^s : (x))/I^{r-1}J^s \simeq (0:x)$. □

4. Minkowski's equality for multiplicity of ideals in one-dimensional local rings

In this section, we shall prove that in a one-dimensional local ring $(R, \mathfrak{m})$, $e(IJ) = e(I) + e(J)$ for all $\mathfrak{m}$-primary ideals $I, J \subset R$. Recall that the zeroeth local cohomology module of R with respect to $\mathfrak{m}$ is the ideal

$$H^0_{\mathfrak{m}}(R) = \{x \in R \mid x\mathfrak{m}^n = 0 \text{ , for some } n \in \mathbb{N}\}.$$

LEMMA 4.1. *Let R be a Noetherian local ring of dimension $d \geq 1$ and I be an $\mathfrak{m}$-primary ideal. Set $S = H^0_{\mathfrak{m}}(R)$. Then*

$$e(I, R) = e((I + S)/S, R/S).$$

PROOF. Observe that

$$\ell\left(\frac{R/S}{(I^n+S)/S}\right)=\ell\left(\frac{R}{I^n+S}\right)=\ell\left(\frac{R}{I^n}\right)-\ell\left(\frac{I^n+S}{I^n}\right)$$
$$=\ell\left(\frac{R}{I^n}\right)-\ell\left(\frac{S}{I^n\cap S}\right).$$

Using Artin-Rees lemma, there exists $n_0\in\mathbb{N}$ such that for all n large,

$$I^n\cap S=I^{n-n_0}(I^{n_0}\cap S)\subseteq I^{n-n_0}S=0.$$

Therefore for n large, $\ell(R/(I^n+S))=\ell(R/I^n)-\ell(S)$. Divide by $n^d/d!$ and take limit to get the required result. $\square$

PROPOSITION 4.2. Let $(R,\mathfrak{m})$ be a 1-dimensional local ring and I,J be $\mathfrak{m}$-primary ideals. Then

$$e(IJ)=e(I)+e(J).$$

PROOF. Using Lemma 4.1, we may pass to $R/H^0_{\mathfrak{m}}(R)$ and so assume that R is Cohen-Macaulay (this is true because $\mathfrak{m}$ is not a associated prime of $H^0_{\mathfrak{m}}(R)$). Without loss of generality, we may assume that the residue field $k=R/\mathfrak{m}$ is infinite. Let $x\in I$ and $y\in J$ so that (x) is a minimal reduction of I and (y) is a minimal reduction of J. Then (xy) is a minimal reduction of IJ. Hence

$$e(IJ)=e(xy)=\lim_{n\to\infty}\ell(R/x^ny^n)/n.$$

Consider the exact sequences

$$0\longrightarrow(x^n)/(x^ny^n)\longrightarrow R/(x^ny^n)\longrightarrow R/(x^n)\longrightarrow 0.$$

Since x^n is a nonzerodivisor, $(x^n)/(x^ny^n)\simeq R/(y^n)$ and so it follows that

$$\ell(R/(x^ny^n))=\ell(R/(x^n))+\ell(R/(y^n)).$$

Divide by n and take limit to see that $e(IJ)=e(I)+e(J)$. $\square$

5. Teissier's approach to Minkowski's inequalities

Mixed multiplicities of ideals: Let I and J be $\mathfrak{m}$-primary ideals of a d-dimensional local ring $(R,\mathfrak{m})$. We know that the function $H(r,s)=\ell(R/I^rJ^s)$, for all large r,s, is given by a polynomial

$$P(r,s)=\frac{1}{d!}\left\{e_0(I|J)r^d+\cdots+\binom{d}{i}e_i(I|J)r^{d-i}s^i+\cdots+e_d(I|J)s^d\right\}+\cdots.$$

Here, $e_i(I|J)$ for $i=0,1,\ldots,d$ are positive integers called the *mixed multiplicities of I and J*. We prove some basic properties of mixed multiplicities in the next result.

LEMMA 5.1. *We have* (1) $e_0(I|J)=e(I), e_d(I|J)=e(J)$.
(2) *For all integers $r,s\geq 0$,*

$$e(I^rJ^s)=e(I)r^d+\cdots+e_i(I|J)\binom{d}{i}r^{d-i}s^i+\cdots+e(J)s^d.$$

(3) *For all positive integers p,q, $e_i(I^p|J^q)=e_i(I|J)p^{d-i}q^i$ for all $i=0,1,\ldots,d$.*
(4) *For all $i=0,1,\ldots,d$, we have $e_i(I|I)=e(I)$.*
(5) *If I is a reduction of K and J is a reduction of L, then $e_i(I|J)=e_i(K|L)$ for all $i=0,1,\ldots,d$.*

PROOF. (1) Let $H(r,s) = P(r,s)$ for all $r \geq l$ and $s \geq k$. Then for all $r \geq l$,

$$\ell(R/I^r J^k) = e_0(I|J)\frac{r^d}{d!} + \text{ terms of degree } < d.$$

Let $m \in \mathbb{N}$ such that $I^m \subseteq J^k$. Then $I^{r+m} \subseteq I^r J^k \subseteq I^r$ for all $r \geq 0$. Hence for $r \geq l$, we have

$$\ell(R/I^{r+m}) \geq H(r,k) \geq \ell(R/I^r).$$

Divide by $r^d/d!$ and take limit $r \to \infty$ to see that $e(I) \geq e_0(I|J) \geq e(I)$. Hence $e_0(I|J) = e(I)$. By symmetry, $e(J) = e_d(I|J)$.
(2) We may assume that $r, s \geq 1$. For large n,

$$\begin{aligned}\ell(R/I^{rn}J^{sn}) &= \frac{1}{d!}\sum_{i=0}^{d} e_i(I|J)\binom{d}{i}(rn)^{d-i}(sn)^i + \cdots \\ &= \frac{n^d}{d!}\sum_{i=0}^{d} e_i(I|J)\binom{d}{i} r^{d-i}s^i + \cdots\end{aligned}$$

It follows that for all $r, s \geq 1$, $e(I^r J^s) = \sum_{i=0}^{d} e_i(I|J)\binom{d}{i} r^{d-i}s^i$.
(3) Using the formula for $e(I^r J^s)$ we have

$$\sum_{i=0}^{d}\binom{d}{i} e_i(I|J)(rp)^{d-i}(sq)^i = e(I^{pr}J^{qs}) = \sum_{i=0}^{d}\binom{d}{i} e_i(I^p|J^q) r^{d-i}s^i.$$

Equate the coefficients of $r^{d-i}s^i$ to get $e_i(I^p|J^q) = p^{d-i}q^i e_i(I|J)$ for all $i = 0, \ldots, d$.
(4) Put $I = J$ in (2) to get

$$\sum_{i=0}^{d} e_i(I|I)\binom{d}{i} r^{d-i}s^i = e(I^{r+s}) = e(I)(r+s)^d = \sum_{i=0}^{d}\binom{d}{i} e(I) r^{d-i}s^i.$$

Equate the coefficients of $r^{d-i}s^i$ to see that $e_i(I|I) = e(I)$ for all $i = 0, 1, \ldots, d$.
(5) Since I is a reduction of K and J is a reduction of L, $I^r J^s$ is a reduction of $K^r L^s$ for all $r, s \geq 1$. Hence

$$\sum_{i=0}^{d}\binom{d}{i} e_i(I|J) r^{d-i}s^i = e(I^r J^s) = e(K^r L^s) = \sum_{i=0}^{d}\binom{d}{i} e_i(K|L) r^{d-i}s^i.$$

Equate the coefficients of $r^{d-i}s^i$ to see that $e_i(I|J) = e_i(K|L)$ for all $i = 0, 1, \ldots, d$. □

Teissier's approach to Minkowski's inequality: Put $r = s = 1$ in part (2) of the above lemma to get

$$e(IJ) = e(I) + \binom{d}{1} e_1(I|J) + \cdots + e_i(I|J)\binom{d}{i} + \cdots + e(J).$$

Teissier compared the expansion

$$(e(I)^{1/d} + e(J)^{1/d})^d = e(I) + \cdots + \binom{d}{i} e(I)^{(d-i)/d} e(J)^{i/d} + \cdots + e(J)$$

with the formula for $e(IJ)$ and proposed the following:

CONJECTURE 5.2 (**Teissier's first conjecture**). Let $(R, \mathfrak{m})$ be a d-dimensional Noetherian local ring and I, J be $\mathfrak{m}$-primary ideals. Then for all $i = 0, 1, \ldots, d$,

$$e_i(I|J)^d \leq e(I)^{d-i} e(J)^i.$$

It is clear that Teissier's first conjecture implies Minkowski's inequality. Teissier approached his first conjecture via his second.

CONJECTURE 5.3 (**Teissier's second conjecture**). Let I and J be $\mathfrak{m}$-primary ideals of a Noetherian local ring of dimension $d \geq 2$. Put $e_i = e_i(I|J)$. Then

$$\frac{e_1}{e_0} \leq \frac{e_2}{e_1} \leq \frac{e_3}{e_2} \leq \cdots \leq \frac{e_d}{e_{d-1}}.$$

PROPOSITION 5.4. Conjecture 5.3 implies the Conjecture 5.2.

PROOF. Apply induction on d. The two conjectures are equivalent for $d = 2$. Assume that the conjecture 5.3 for $d-1$ and $d \geq 3$, implies the conjecture 5.2. Then for $i = 0, 1, \ldots, d-1$, $e_i^{d-1} \leq e_0^{d-1-i} e_{d-1}^i$. Note that

$$\frac{e_1}{e_0}\frac{e_2}{e_1}\cdots\frac{e_i}{e_{i-1}} = \frac{e_i}{e_0} \leq \left(\frac{e_d}{e_{d-1}}\right)^i.$$

Hence

$$e_i^d \leq e_0^{d-1-i} e_i e_{d-1}^i = e_0^{d-i} e_{d-1}^i \frac{e_i}{e_0} \leq e_0^{d-i} e_{d-1}^i \left(\frac{e_d}{e_{d-1}}\right)^i = e_0^{d-i} e_d^i.$$

□

6. Rees-Sharp proof of Minkowski's inequalities

In this section, we shall prove Minkowski's inequality for multiplicities in Noetherian local rings. We shall use Lech's formula for the multiplicity of an ideal generated by a system of parameters. Let $a_1, a_2, \ldots, a_d$ be a system of parameters in a d-dimensional local ring and $I = (a_1, a_2, \ldots, a_d)$. Then the Lech's formula for $e(I)$ is:

$$e(I) = \lim_{n\to\infty} \frac{\ell(R/(a_1^n, a_2^n, \ldots, a_d^n))}{n^d}.$$

For $n \in \mathbb{N}$, we write $I^{[n]} = (a_1^n, a_2^n, \ldots, a_d^n)$.

LEMMA 6.1. *Let $(R, \mathfrak{m})$ be a two-dimensional Noetherian local ring, $I = (a, b)$ and J be $\mathfrak{m}$-primary ideals. Then for all $n \geq 1$,*

$$\ell(R/(J^n : I^{[n]})) \leq \ell(R/I^{[n]}) + 2\ell(R/J^n) - \ell(R/I^n J^n).$$

PROOF. Use the following diagram

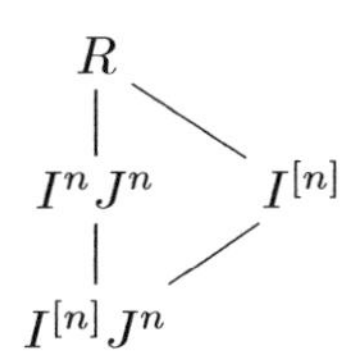

to see that

$$\ell(R/I^n J^n) + \ell(I^n J^n / I^{[n]} J^n) = \ell(R/I^{[n]}) + \ell(I^{[n]}/I^{[n]} J^n).$$

In order to estimate $\ell(I^{[n]}/I^{[n]} J^n)$, consider the complex of R-modules

$$0 \longrightarrow R \xrightarrow{f} R/J^n \oplus R/J^n \xrightarrow{g} \frac{(a^n, b^n)}{(a^n J^n + b^n J^n)} \longrightarrow 0$$

where $f(r) = (r[b^n], -r[a^n])$ and $g([r],[s]) = [ra^n + sb^n]$ for all $r, s \in R$. Then $\ker f = (J^n : I^{[n]})$ and hence the image of f is isomorphic to $R/(J^n : I^{[n]})$. Therefore

$$\ell(I^{[n]}/I^{[n]}J^n) \leq 2\ell(R/J^n) - \ell(R/J^n : I^{[n]})$$

and hence $\ell(R/J^n : I^{[n]}) \leq \ell(R/I^{[n]}) - \ell(R/I^{[n]}J^n) + 2\ell(R/J^n)$. □

PROPOSITION 6.2. Let $(R, \mathfrak{m})$ be a 2-dimensional local ring and I, J be $\mathfrak{m}$-primary ideals. Then

$$e(IJ) \leq 2e(I) + 2e(J).$$

PROOF. We may assume without loss of generality that $R/\mathfrak{m}$ is infinite. Let (a, b) be a minimal reduction of I. Then $(a, b)J$ is a reduction of IJ. Hence $e(IJ) = e((a, b)J)$ and $e(I) = e(a, b)$. Therefore we may assume that $I = (a, b)$. By Lemma 6.1, we have

$$\begin{aligned}\ell(R/I^nJ^n) &\leq \ell(R/I^{[n]}) + 2\ell(R/J^n) - \ell(R/J^n : I^{[n]})\\ &\leq \ell(R/I^{[n]}) + 2\ell(R/J^n).\end{aligned}$$

Divide by $n^2/2$ and take the limit $n \to \infty$. This gives $e(IJ) \leq 2e(I) + 2e(J)$. □

THEOREM 6.3. *Let $(R, \mathfrak{m})$ be a 2-dimensional local ring and I, J be $\mathfrak{m}$-primary ideals. Then*

$$e_1(I|J)^2 \leq e(I)e(J).$$

PROOF. Using Proposition 6.2 for I^r and J^s, we get

$$\begin{aligned}r^2e(I) + 2e_1(I|J)rs + s^2e(J) = e(I^rJ^s) &\leq 2e(I^r) + 2e(J^s)\\ &= 2r^2e(I) + 2s^2e(J).\end{aligned}$$

Therefore for all $r, s \in \mathbb{N}$, we have

$$f(r, s) := r^2e(I) - 2rse_1(I|J) + e(J)s^2 \geq 0.$$

Hence the discriminant of $f(r, s)$, namely, $4e_1(I|J)^2 - 4e(I)e(J) \leq 0$. This implies that $e_1(I|J)^2 \leq e(I)e(J)$. □

THEOREM 6.4 (**Rees-Sharp**). *Let $(R, \mathfrak{m})$ be a local ring of dimension $d \geq 2$. Let I and J be $\mathfrak{m}$-primary ideals. Then Teissier's second conjecture is true.*

PROOF. Apply induction on d. We have established the conjecture for $d = 2$. Let $d \geq 3$. Let $a \in I$ be superficial for I, J and the set of minimal primes of R. Then for $i = 0, 1, \ldots, d-1$,

$$e_i(I/(a) \mid J + (a)/(a)) = e_i(I|J).$$

Hence

$$\frac{e_1}{e_0} \leq \frac{e_2}{e_1} \leq \frac{e_3}{e_2} \leq \cdots \leq \frac{e_{d-1}}{e_{d-2}}.$$

If we use a superficial element from J, we obtain the remaining inequality by symmetry. □

7. Complete ideals and discrete valuation rings

DEFINITION 7.1. A local domain $(S, \mathfrak{n})$ is said to dominate a local domain $(R, \mathfrak{m})$ birationally if $R \subset S \subset K$ where K is the fraction field of R and $\mathfrak{n} \cap R = \mathfrak{m}$. If S birationally dominates R, we write $S \succ R$ or $R \prec S$.

PROPOSITION 7.2. Let $(R, \mathfrak{m})$ be a local domain of positive dimension. Then there is a discrete valuation ring $(V, \mathfrak{n})$ birationally dominating $(R, \mathfrak{m})$.

PROOF. First, we show that there exists an $x \in \mathfrak{m}$ such that $x^k \notin \mathfrak{m}^{k+1}$ for all $k \geq 1$. Let $\mathfrak{m} = (x_1, x_2, \dots, x_n)$ and assume by way of contradiction that $x_i^{k_i} \in \mathfrak{m}^{k_i+1}$ for some k_i and for all $i = 1, 2, \dots, n$. Let $k = \max k_i$. Since $\mathbf{x}^{[k]} := (x_1^k, x_2^k, \dots, x_n^k)$ is a reduction of $\mathfrak{m}^k$, there exists an s such that $\mathbf{x}^{[k]}\mathfrak{m}^{ks} = \mathfrak{m}^{ks+k}$. Hence $\mathfrak{m}^{ks+k} \subset \mathfrak{m}^{sk+k+1}$ which yields $\mathfrak{m}^{ks+k} = 0$. This is a contradiction as $\dim R \geq 1$. Thus we may assume without loss of generality that for $x_1 = x$, $x^k \notin \mathfrak{m}^{k+1}$ for all k.

The ring $S = R[\mathfrak{m}/x] = R[x_2/x, x_3/x, \dots, x_n/x]$ is called a *monoidal transform* of R. It is easy to see that $S = \{b/x^k : b \in \mathfrak{m}^k \text{ for some } k\}$. The ideal $xS = \mathfrak{m}S$ is a proper ideal. Indeed, if $1 \in xS$ then $1 = bx/x^d$ for some $d \geq 1$ and $b \in \mathfrak{m}^d$. Hence $x^d \in \mathfrak{m}^{d+1}$, contradicting the choice of x. Thus xS is a height one ideal of S. Let Q be a minimal prime of xS. By Krull-Akizuki theorem, the integral closure T of S_Q in its fraction field K is a one dimensional Noetherian domain. Let N be a maximal ideal of T contracting to the maximal ideal of S_Q. Then $NT_N \cap R = \mathfrak{m}$ and hence T_N is the desired discrete valuation ring birationally dominating R. $\square$

THEOREM 7.3 (**Lipman's theorem, [17]**). *Let S be a Noetherian domain with fraction field K and let I be a proper ideal of S. Then*

$$\overline{I} = \bigcap_V IV \cap S$$

where the intersection is over all discrete valuation rings V in K such that $V \supset S$.

PROOF. Since principal ideals in integrally closed domains are complete and intersections of complete ideals are complete, the ideal J on the right hand side of the above equation is complete. Hence $\overline{I} \subseteq J$. Conversely let $x \notin \overline{I}$. Then we find a discrete valuation ring $V \supset S$ in K such that $x \notin IV$. Put $T = S[Ix^{-1}]$. Then $x^{-1}IT$ is a proper ideal of T. Indeed, if $x^{-1}IT = T$, then $1 = a_1/x + a_2/x^2 + \cdots + a_n/x^n$, where $a_i \in I^i$ for $i = 1, 2, \dots, n$. Hence $x^n = a_1x^{n-1} + a_2x^{n-2} + \cdots + a_n$ which shows that $x \in \overline{I}$. This is a contradiction. Pick a minimal prime Q of $x^{-1}IT$. By Proposition 7.2, there exists a discrete valuation ring $(V, \mathfrak{n})$ such that $V \succ T_Q$. Hence $x^{-1}IT \subset Q \subset QT_Q = \mathfrak{n} \cap T_Q$ and $x^{-1}IV \subseteq \mathfrak{n}$. Thus $x \notin IV$. $\square$

8. Minkowski equality in dimension two

If I and J are $\mathfrak{m}$-primary ideals of a local ring $(R, \mathfrak{m})$ and J is a reduction of I then $e(I) = e(J)$. Moreover, a deep theorem of Rees asserts that in case R is quasi-unmixed and $J \subset I$ with $e(I) = e(J)$, then $\overline{I} = \overline{J}$. It is natural to ask for a numerical criterion for $\overline{I} = \overline{J}$ if there is no containment relation among I and J. We shall see that if R is quasi-unmixed and $e(I) = e_i(I|J)$ for all $i = 1, \dots, d$, then $\overline{I} = \overline{J}$. We shall prove this in dimension 2 in this section.

PROPOSITION 8.1. Let R be a d-dimensional Noetherian local ring and I, J be $\mathfrak{m}$-primary ideals of R. Suppose there exist positive integers r, s such that $\overline{I^r} = \overline{J^s}$. Then

$$\frac{e_1}{e_0} = \frac{e_2}{e_1} = \cdots = \frac{e_d}{e_{d-1}} = \frac{r}{s}.$$

PROOF. Let $\overline{I^r} = \overline{J^s}$. Then for all $i = 0, 1, \ldots, d$,

$$r^{d-i}s^i e_i = e_i(\overline{I^r}|\overline{J^s}) = e_i(\overline{I^r}|\overline{I^r}) = r^d e(I).$$

This implies that $e_i/e_{i-1} = r/s$ for all $i = 1, \ldots, d$. □

Rees and Sharp proved that the converse of the above theorem is true for two-dimensional quasi-unmixed local rings. They used the theory of degree functions and $\mathfrak{m}$-valuations. Let $(R, \mathfrak{m})$ be a Noetherian local ring of dimension d. Let $k(\mathfrak{p})$ denote the field of fractions of $R/\mathfrak{p}$ where $\mathfrak{p}$ is a prime ideal of R and let $\mathbb{Z}$ denote the additive group of integers.

DEFINITION 8.2. An $\mathfrak{m}$-valuation of a Noetherian local ring $(R, \mathfrak{m})$ is a map $v : R \to \mathbb{Z} \cup \{\infty\}$ which is a composition of the natural ring homomorphism $R \to R/\mathfrak{p}$, for a minimal prime $\mathfrak{p}$ of dimension d and a valuation $v : k(\mathfrak{p}) \to \mathbb{Z} \cup \{\infty\}$ such that (1) $v(x) \geq 0$ for all $x \in R/\mathfrak{p}$, (2) $v(x) > 0$ for all $x \in \mathfrak{m}/\mathfrak{p}$ and, (3) the residue field K_v of v is a finitely generated extension of $R/\mathfrak{m}$ of transcendence degree $d - 1$.

COROLLARY 8.3 ([**17**, Proposition 1.1]). Let R be a complete, universally catenary, Noetherian local domain of dimension d. Let I be a proper ideal of R. Then for every $x \notin \overline{I}$, there exists an $\mathfrak{m}$-valuation v of R such that $v(x) < v(I)$.

PROOF. Let $x \notin \overline{I}$. As in the proof of Proposition 7.2 and Theorem 7.3, there exists a discrete valuation domain T_N birationally dominating R such that $x \notin IT_N$ and NT_N contracts to the maximal ideal $\mathfrak{m}$ of R. It is sufficient to show that $\operatorname{tr deg}_{k(m)} k(NT_N) = d - 1$. Since R is a complete local domain, it is excellent and hence T is finitely generated over R. As R is universally catenary, using dimension formula,

$$\operatorname{tr deg}_{k(m)} k(NT_N) = \operatorname{ht} \mathfrak{m} + \operatorname{tr deg}_R T - \operatorname{ht} N = d - 1.$$

□

LEMMA 8.4. *Let $(R, \mathfrak{m})$ be a 2-dimensional quasi-unmixed local ring. Let v be an $\mathfrak{m}$-valuation of R. For a positive integer i, let $c(v)_i = \{a \in R \mid v(a) \geq i\}$. Then there exists a positive real number λ such that for all positive integers i,*

$$\ell\left(\frac{R}{c(v)_i}\right) \geq \lambda i^2.$$

PROOF. Let $(V, \mathfrak{n})$ be the discrete valuation ring corresponding to the valuation v. Let $\mathfrak{p}$ be the minimal prime ideal of R such that $R/\mathfrak{p} \subseteq V$. Write $(\_)'$ to denote images in $R/\mathfrak{p}$. There exist $x_1, x_2 \in R \setminus \mathfrak{p}$ such that x_1'/x_2' in $V/\mathfrak{n}$ is transcendental over $R/\mathfrak{m}$. This implies that $v(x_1) = v(x_2) = n$, say, for some positive integer n. Observe that $n > 0$. If not, then $v(x_1) = v(x_2) = 0$ implies that $x_1', x_2' \in V \setminus \mathfrak{n}$. Therefore $x_1'/x_2' \in R/\mathfrak{m}$ giving a contradiction.

Let r be a positive integer. By definition, $\mathfrak{m}c(v)_{nr} \subseteq c(v)_{nr+1}$ implying that $c(v)_{nr}/c(v)_{nr+1}$ is an $R/\mathfrak{m}$-vector space. We claim that $x_1^r, x_1^{r-1}x_2, \ldots, x_2^r$ are linearly independent over $R/\mathfrak{m}$. If not, then there exist $a_1, \ldots, a_{r+1} \in R/\mathfrak{m}$, not all zero, such that

$$a_1x_1^r + a_2x_1^{r-1}x_2 + \cdots + a_{r+1}x_2^r = 0$$

in the vector space. Dividing by x_2^r, we get x_1/x_2 is algebraic over $R/\mathfrak{m}$, a contradiction. Since $v(x_1^{r-i}x_2^i) = nr$ for all $i = 0, 1, \dots, r$, it follows that

$$\dim_{R/\mathfrak{m}}(c(v)_{nr}/c(v)_{nr+1}) \geq r + 1.$$

Let i be a positive integer and s be the greatest integer such that $sn \leq i - 1$. Then

$$\begin{aligned} \ell\left(\frac{R}{c(v)_i}\right) &\geq \ell\left(\frac{R}{c(v)_{sn+1}}\right) \\ &\geq \ell\left(\frac{R}{c(v)_1}\right) + \ell\left(\frac{c(v)_n}{c(v)_{n+1}}\right) + \cdots + \ell\left(\frac{c(v)_{sn}}{c(v)_{sn+1}}\right) \\ &\geq 1 + 2 + \cdots + (s+1) = \frac{(s+1)(s+2)}{2} \geq \frac{i^2}{2n^2} \end{aligned}$$

as $(s+1)n \geq i$. The result follows upon taking $\lambda = 1/2n^2$. □

THEOREM 8.5. *Let $(R, \mathfrak{m})$ be a 2-dimensional quasi-unmixed local ring with infinite residue field. Let I, J be $\mathfrak{m}$-primary ideals of R. Suppose r and s are positive integers such that $r^2e(I) = s^2e(J) = rse_1(I|J)$. Then I^r and J^s have the same integral closure.*

PROOF. We may assume that R is a complete local ring with infinite residue field. It is sufficient to show that if $e(I) = e(J) = e_1(I|J)$, then $\overline{I} = \overline{J}$. We may assume that $I = (a_1, a_2)$ is generated by system of parameters. Using Lemma 6.1

$$\ell\left(\frac{R}{J^n : (a_1^n, a_2^n)}\right) \leq \ell\left(\frac{R}{(a_1^n, a_2^n)}\right) + 2\ell\left(\frac{R}{J^n}\right) - \ell\left(\frac{R}{I^nJ^n}\right).$$

Taking limit as $n \to \infty$ after dividing by $n^2/2$ and using Lech's formula, it follows that

$$\lim_{n\to\infty} \ell\left(\frac{R}{J^n : (a_1^n, a_2^n)}\right) \cdot \frac{1}{n^2} = 0.$$

We show that both a_1 and a_2 are integral over J. Suppose a_1 is not integral over J. Then there exists a minimal prime $\mathfrak{p}$ of R so that a_1 is not integral over $J + \mathfrak{p}/\mathfrak{p}$. Now $R/\mathfrak{p}$ is a complete local domain of dimension 2. Hence by Corollary 8.3, there exists an $\mathfrak{m}$-valuation v of $R/\mathfrak{p}$ so that $v(a_1) < v(J)$. Set $j = v(J) - v(a_1) > 0$. Let $b \in J^n : (a_1^n, a_2^n)$, then $ba_1^n \in J^n$ and hence $v(b) + nv(a_1) \geq nv(J)$. This implies that $v(b) \geq nj$ and using the notation as in Lemma 8.4, we get $b \in c(v)_{nj}$. Therefore for all $n > 0$, $(J^n : (a_1^n, a_2^n)) \subseteq c(v)_{nj}$ and hence using Lemma 8.4,

$$\ell\left(\frac{R}{J^n : (a_1^n, a_2^n)}\right) \geq \ell\left(\frac{R}{c(v)_{nj}}\right) \geq \lambda(nj)^2 > 0.$$

Thus

$$\lim_{n\to\infty} \ell\left(\frac{R}{J^n : (a_1^n, a_2^n)}\right) \cdot \frac{1}{n^2} \geq \lambda j^2 > 0$$

giving a contradiction. So both a_1 and a_2 are integral over J and hence $I \subseteq \overline{J}$. By symmetry, $J \subseteq \overline{I}$, completing the proof. □

9. Minkowski's equality in dimension $d \geq 3$

The proof of Minkowski equality in dimension 3 or higher, was reduced to dimension 2 by D. Katz [**15**] by passing to certain subrings of the total quotient ring of R having smaller dimension than $\dim R$. In this section, we present an alternative proof using the specialization property of the integral closure of an ideal first proved by Shiroh Itoh [**14**, Theorem 1] if R is Cohen-Macaulay of dimension $d \geq 2$ and is analytically unramified. Hong and Ulrich [**11**] provided another proof for analytically unramified universally catenary local rings.

We use the notion of general extensions in the proof. Let $x_1, \dots, x_n$ be a set of indeterminates over R. The ring $S = R[x_1, \dots, x_n]_{\mathfrak{m}R[x_1,\dots,x_n]}$ is called a *general extension* of the ring R. Let $I = (a_1, \dots, a_n)$ be an ideal of R. The element $x = \sum_{i=1}^{n} a_i x_i$ is called a *general element* of I. Then S/R is a faithfully flat extension, $\dim R = \dim S$ and $e(J, R) = e(JS, S)$ for any $\mathfrak{m}$-primary ideal J of R. For proof of these statements and other properties of the ring S, the reader may refer to [**25**].

THEOREM 9.1 ([**25**, Lemma 2.4]). *Let $(R, \mathfrak{m})$ be a Noetherian local ring and $I = (a_1, \dots, a_n), J$ be ideals of R. Let $S = R[x_1, \dots, x_n]_{\mathfrak{m}R[x_1,\dots,x_n]}$ be a general extension of R. Then $x = a_1x_1 + \cdots + a_nx_n$ is Rees superficial for the pair (IS, JS).*

We are now ready to give the proof of the main theorem.

THEOREM 9.2. *Let $(R, \mathfrak{m})$ be a d-dimensional quasi-unmixed local ring with infinite residue field and $d \geq 2$. Let I, J be $\mathfrak{m}$-primary ideals. Then the following are equivalent:*
(1) $e(IJ)^{1/d} = e(I)^{1/d} + e(J)^{1/d}$.
(2) *There exists positive integers r, s such that*

$$\frac{e_1(I|J)}{e_0(I|J)} = \frac{e_2(I|J)}{e_1(I|J)} = \cdots = \frac{e_d(I|J)}{e_{d-1}(I|J)} = \frac{r}{s}.$$

(3) *There exists positive integers r, s such that $\overline{I^r} = \overline{J^s}$.*

PROOF. **(1) $\Rightarrow$ (2)**: Consider

$$e(IJ) = (e(I)^{1/d} + e(J)^{1/d})^d = \sum_{i=0}^{d} \binom{d}{i} e(I)^{(d-i)/d} e(J)^{i/d}.$$

Since $e(IJ) = \sum_{i=0}^{d} \binom{d}{i} e_i$, we get

$$\sum_{i=0}^{d} \binom{d}{i} \left(e(I)^{(d-i)/d} e(J)^{i/d} - e_i \right) = 0.$$

Using the inequality $e_i^d \leq e(I)^{d-i} e(J)^i$, it follows that $e_i = e(I)^{(d-i)/d} e(J)^{i/d}$ for all $i = 0, \dots, d$. For $i = 1, \dots, d$, consider

$$\frac{e_i}{e_{i-1}} = \frac{e(I)^{(d-i)/d} e(J)^{i/d}}{e(I)^{(d-i+1)/d} e(J)^{(i-1)/d}} = \left(\frac{e(J)}{e(I)} \right)^{1/d}.$$

Hence $\frac{e_i}{e_{i-1}} = \frac{e_1}{e_0} = \frac{r}{s}$, for all $i = 1, 2 \dots, d$.

(2) $\Rightarrow$ (1): Note that it is sufficient to show that $e_i = e(I)^{(d-i)/d} e(J)^{i/d}$, for all $i = 0, 1, \ldots, d$. Observe that

$$\left(\frac{e_i}{e_0}\right)^{d-i} = \left(\frac{e_i}{e_{i-1}}\right)^{d-i} \cdots \left(\frac{e_1}{e_0}\right)^{d-i}$$
$$= \left(\frac{e_d}{e_{d-1}} \cdots \frac{e_{i+1}}{e_i}\right) \cdots \left(\frac{e_d}{e_{d-1}} \cdots \frac{e_{i+1}}{e_i}\right) = \left(\frac{e_d}{e_i}\right)^i.$$

To prove the equivalence of (2) and (3) it is sufficient to show that $e_0 = e_1 = \cdots = e_d$ if and only if $\overline{I} = \overline{J}$. As $e_i(I^r|J^s) = r^{d-i}s^i e_i(I|J)$, we may replace I^r by I and J^s by J.

(3) $\Rightarrow$ (2) : We have proved it as Proposition 8.1.

Observe that the quasi-unmixed property of the ring was not used in proving the above equivalences. We however need it to prove the following:

(2) $\Rightarrow$ (3) : We proceed by induction on d. The case $d = 2$ is proved in Theorem 8.5. Suppose $d > 2$.

We may assume that R is complete and $I = (a_1, \ldots, a_d)$ is generated by system of parameters. Let $Z_1, \ldots, Z_d$ be indeterminates over R and put $S = R[Z_1, \ldots, Z_d]_{\mathfrak{m}R[Z_1,\ldots,Z_d]}$. Let $x = \sum_{i=1}^d a_i Z_i \in S$. Using Theorem 9.1, it follows that there exists $c > 0$ such that for all $r \geq c$ and $s \geq 0$,

$$I^r J^r S \cap (x)S = xI^{r-1}J^s S.$$

We now prove that $S/(x)$ is quasi-unmixed. By [**18**, Theorem 31.6], R is universally catenary and hence S is catenary. Since $\dim S/(x) = d - 1$, any minimal prime of (x) in S has height 1. By the catenary property of S, it follows that $\dim(S/\mathfrak{p}) = \dim(S/(x)) = d - 1$ for all minimal primes $\mathfrak{p}$ of (x) in S. Therefore $S/(x)$ is quasi-unmixed by [**18**, Theorem 31.6]. Since $(0 :_S x) \subseteq (0 :_S IS)$, using Theorem 3.8 it follows that

$$\ell\left(\frac{S}{I^r J^s S}\right) - \ell\left(\frac{S}{I^{r-1}J^s S}\right) = \ell\left(\frac{S}{I^r J^s S + (x)S}\right) - \ell((0 :_S x)).$$

Let $e'_0, \ldots, e'_d$ denote the mixed multiplicities of I and J as ideals in $S/(x)$. Then using above equation, $e_0 = e'_0, \ldots, e_{d-1} = e'_{d-1}$. As $\dim S/(x)S = d - 1$, using induction hypothesis we get $J(S/(x)) \subseteq \overline{I(S/(x))}$. Let $\mathfrak{p}$ be a minimal prime of R and $\tilde{R} = R/\mathfrak{p}$. Consider the natural surjective map

$$\frac{S}{(x)} = \frac{R[Z_1, \ldots, Z_d]_{\mathfrak{m}R[z_1,\ldots,Z_d]}}{(x)} \to T = \frac{\tilde{R}[Z_1, \ldots, Z_d]_{\mathfrak{m}\tilde{R}[z_1,\ldots,Z_d]}}{(x)} \to 0.$$

Using the persistence property of integral closure, it follows that $JT \subseteq \overline{IT}$. As $\tilde{R}$ is a complete local domain, it is analytically unramified and hence $JT \subseteq \overline{IT} = \overline{I}T$ by [**11**, Corollary 2.3]. Therefore,

$$J\tilde{R} \subseteq JT \cap \tilde{R} \subseteq \overline{I\tilde{R}}.$$

Since this is true for all minimal prime ideals $\mathfrak{p}$ of R, it follows that $J \subseteq \overline{I}$. By symmetry $I \subseteq \overline{J}$, completing the proof. □

We now present an algebraic proof of the Rees multiplicity theorem using Minkowski's equality and a geometric proof in dimension 2 using negative definiteness of the intersection form. The algebraic proof is given by Teissier [**27**].

THEOREM 9.3 (**Rees**, [**23**]). *Let $(R, \mathfrak{m})$ be a d-dimensional quasi-unmixed local ring and $J \subseteq I$ be $\mathfrak{m}$-primary ideals. If $e(I) = e(J)$, then $\overline{I} = \overline{J}$.*

PROOF. Note that $e(IJ) \geq e(I^2) = 2^d e(I)$. On the other hand,

$$e(IJ) = \sum_{i=0}^{d} \binom{d}{i} e_i(I|J) \leq \sum_{i=0}^{d} \binom{d}{i} e(I)^{\frac{d-i}{d}} e(J)^{\frac{i}{d}} = 2^d e(I).$$

Hence $e_i(I|J) = e(I)$ for all $i = 0, \ldots, d$. Since R is quasi-unmixed, it follows that $\overline{I} = \overline{J}$. □

THEOREM 9.4. *Let R be a geometric local domain of dimension 2, $I \subset J$ be ideals primary for the maximal ideal of R. If $e(I) = e(J)$, then $\overline{I} = \overline{J}$.*

PROOF. Let $f : X \to \operatorname{Spec} R$ be a resolution of singularities such that f^*I, f^*J are locally invertible sheaves of ideals in X. They define divisors D, E in X such that $E \leq D$. By [**22**], $e(I) = -D^2$, $e(J) = -E^2$.

Write $D = E + E'$ for an effective divisor E'. One property of D and E is that $D.D_i \leq 0$ for every irreducible component D_i of support D with strict inequality for some i (similarly for E), and every D_i occurs in both D, E.

Now $(E + E')^2 = E^2 + 2E.E' + E'^2$. The term $E.E' \leq 0$ by the remark above. If E' is non-zero, then $E'^2 < 0$ by negative definiteness of the intersection form. It follows that if D is strictly bigger than E, then $D^2 < E^2$, i.e., $e(I) > e(J)$. By assumption, $e(I) = e(J)$. Hence $D = E$. Since $\overline{I} \subset \overline{J}$, the proof shows that $\overline{I} = \overline{J}$. □

We end with an example which illustrates that Theorem 9.2 fails if the ring is not quasi-unmixed.

EXAMPLE 9.5. Let $R = k[[X, Y, Z]]/(XY, XZ)$ where k is an infinite field and let x, y, z denote the images of X, Y and Z in R respectively. Observe that R is a 2-dimensional Noetherian local ring but it is not quasi-unmixed as the minimal primes of R are (x) and (y, z). Set $\mathfrak{m} = (x, y, z)$ and $I = (y, x^2 + z)$. We first show that $e(I) = e(\mathfrak{m})$. Using the associativity formula, we get

$$e(I, R) = e\left(\frac{I + (x)}{(x)}, \frac{R}{(x)}\right) \ell(R_{(x)}) = e((y, z), k[[y, z]]) = 1$$

and

$$e(\mathfrak{m}, R) = e\left(\frac{\mathfrak{m} + (x)}{(x)}, \frac{R}{(x)}\right) \ell(R_{(x)}) = e((y, z), k[[y, z]]) = 1.$$

As in the proof of Theorem 9.3, one can now conclude that $e(I) = e_1(I|\mathfrak{m}) = e(\mathfrak{m})$.

We claim that $\overline{I} \neq \overline{\mathfrak{m}}$. As z satisfies the equation $z^2 - z(x^2 + z) = 0$, it follows that $z \in \overline{I}$ and hence $(x^2, y, z) \subseteq \overline{I}$. In order to prove the claim, it is now sufficient to show that $x \notin \overline{I}$. If x is integral over I, then it is integral over $I(R/(y, z))$ in the ring $R/(y, z)$. In other words, x is integral over (x^2) in $k[[x]]$. This gives a contradiction as (x^2) is integrally closed in $k[[x]]$. Hence the claim is true.

Acknowledgements: Thanks are due to D. Katz and B. Teissier for a careful reading of the manuscript and for providing detailed comments which improved the exposition. The third author thanks Aldo Conca and Marilina Rossi for inviting him to Genoa to offer a course on Hilbert functions. Some of the topics covered in this paper were presented in this course. Financial support from Istituto Nazionale di Alta Matematica made this visit possible.

References

[1] S. S. Abhyankar, *Algebraic space curves*, Les Presses de l'Université de Montréal, Montreal, Que., 1971. Séminaire de Mathématiques Supérieures, No. 43 (Été 1970). MR0399109

[2] S. S. Abhyankar, *Resolution of singularities of embedded algebraic surfaces*, Pure and Applied Mathematics, Vol. 24, Academic Press, New York-London, 1966. MR0217069

[3] P. B. Bhattacharya, *The Hilbert function of two ideals*, Proc. Cambridge Philos. Soc. **53** (1957), 568–575. MR0089835

[4] R. Bondil, *Geometry of superficial elements* (English, with English and French summaries), Ann. Fac. Sci. Toulouse Math. (6) **14** (2005), no. 2, 185–200. MR2141180

[5] J. Briançon and J.-P. Speder, *La trivialité topologique n'implique pas les conditions de Whitney* (French, with English summary), C. R. Acad. Sci. Paris Sér. A-B **280** (1975), no. 6, Aiii, A365–A367. MR0425165

[6] S. D. Cutkosky, *Asymptotic multiplicities of graded families of ideals and linear series*, Adv. Math. **264** (2014), 55–113, DOI 10.1016/j.aim.2014.07.004. MR3250280

[7] Steven Dale Cutkosky, Parangama Sarkar, and Hema Srinivasan, *M*ixed multiplicities of filtrations, Trans. Amer. Math. Soc., electronically published on January 16, 2019, DOI: http://doi.org/10.1090/tran/7745 (to appear in print).

[8] L. Ein, R. Lazarsfeld, and K. E. Smith, *Uniform approximation of Abhyankar valuation ideals in smooth function fields*, Amer. J. Math. **125** (2003), no. 2, 409–440. MR1963690

[9] D. Eisenbud and H. I. Levine, *An algebraic formula for the degree of a C^∞ map germ*, Ann. of Math. (2) **106** (1977), no. 1, 19–44, DOI 10.2307/1971156. With an appendix by Bernard Teissier, "Sur une inégalité à la Minkowski pour les multiplicités". MR0467800

[10] W. Fulton, *Intersection theory*, Ergebnisse der Mathematik und ihrer Grenzgebiete (3) [Results in Mathematics and Related Areas (3)], vol. 2, Springer-Verlag, Berlin, 1984. MR732620

[11] J. Hong and B. Ulrich, *Specialization and integral closure*, J. Lond. Math. Soc. (2) **90** (2014), no. 3, 861–878, DOI 10.1112/jlms/jdu053. MR3291804

[12] J. Huh, *Milnor numbers of projective hypersurfaces and the chromatic polynomial of graphs*, J. Amer. Math. Soc. **25** (2012), no. 3, 907–927, DOI 10.1090/S0894-0347-2012-00731-0. MR2904577

[13] C. Huneke and I. Swanson, *Integral closure of ideals, rings, and modules*, London Mathematical Society Lecture Note Series, vol. 336, Cambridge University Press, Cambridge, 2006. MR2266432

[14] S. Itoh, *Coefficients of normal Hilbert polynomials*, J. Algebra **150** (1992), no. 1, 101–117, DOI 10.1016/S0021-8693(05)80052-3. MR1174891

[15] D. Katz, *Note on multiplicity*, Proc. Amer. Math. Soc. **104** (1988), no. 4, 1021–1026, DOI 10.2307/2047584. MR929434

[16] J. Lipman, *Desingularization of two-dimensional schemes*, Ann. Math. (2) **107** (1978), no. 1, 151–207, DOI 10.2307/1971141. MR0491722

[17] J. Lipman, *On complete ideals in regular local rings*, Algebraic geometry and commutative algebra, Vol. I, Kinokuniya, Tokyo, 1988, pp. 203–231. MR977761

[18] H. Matsumura, *Commutative ring theory*, 2nd ed., Cambridge Studies in Advanced Mathematics, vol. 8, Cambridge University Press, Cambridge, 1989. Translated from the Japanese by M. Reid. MR1011461

[19] J. Milnor, *Singular points of complex hypersurfaces*, Annals of Mathematics Studies, No. 61, Princeton University Press, Princeton, N.J.; University of Tokyo Press, Tokyo, 1968. MR0239612

[20] M. Mustaţă, *On multiplicities of graded sequences of ideals*, J. Algebra **256** (2002), no. 1, 229–249, DOI 10.1016/S0021-8693(02)00112-6. MR1936888

[21] Patrick Popescu-Pampu, How Teissier mixed multiplicities, Singular Landscapes, 2015.

[22] C. P. Ramanujam, *On a geometric interpretation of multiplicity*, Invent. Math. **22** (1973/74), 63–67, DOI 10.1007/BF01425574. MR0354663

[23] D. Rees, $\mathfrak{a}$-*transforms of local rings and a theorem on multiplicities of ideals*, Proc. Cambridge Philos. Soc. **57** (1961), 8–17, DOI 10.1017/s0305004100034800. MR0118750

[24] D. Rees, *Generalizations of reductions and mixed multiplicities*, J. London Math. Soc. (2) **29** (1984), no. 3, 397–414, DOI 10.1112/jlms/s2-29.3.397. MR754926

[25] D. Rees, *The general extension of a local ring and mixed multiplicities*, Algebra, algebraic topology and their interactions (Stockholm, 1983), Lecture Notes in Math., vol. 1183, Springer, Berlin, 1986, pp. 339–360, DOI 10.1007/BFb0075469. MR846458

[26] D. Rees and R. Y. Sharp, *On a theorem of B. Teissier on multiplicities of ideals in local rings*, J. London Math. Soc. (2) **18** (1978), no. 3, 449–463, DOI 10.1112/jlms/s2-18.3.449. MR518229

[27] B. Teissier, *On a Minkowski-type inequality for multiplicities. II*, C. P. Ramanujam—a tribute, Tata Inst. Fund. Res. Studies in Math., vol. 8, Springer, Berlin-New York, 1978, pp. 347–361. MR541030

[28] B. Teissier, *Cycles évanescents, sections planes et conditions de Whitney* (French), Singularités à Cargèse (Rencontre Singularités Géom. Anal., Inst. Études Sci., Cargèse, 1972), Soc. Math. France, Paris, 1973, pp. 285–362. Astérisque, Nos. 7 et 8. MR0374482

[29] Lê Dũng Tráng, *Un critère d'équisingularité* (French), Singularités à Cargèse (Rencontre Singularités Géom. Anal., Inst. Études Sci., Cargèse, 1972), Soc. Math. France, Paris, 1973, pp. 183–192. Astérisque, Nos. 7 et 8. MR0361148

[30] O. Zariski, *Some open questions in the theory of singularities*, Bull. Amer. Math. Soc. **77** (1971), 481–491, DOI 10.1090/S0002-9904-1971-12729-5. MR0277533

Indian Institute of Technology Bombay, Mumbai, INDIA 400076
Email address: `kriti@math.iitb.ac.in`

Indian Institute of Technology Bombay, Mumbai, INDIA 400076
Email address: `gurjar@math.iitb.ac.in`

Indian Institute of Technology Bombay, Mumbai, INDIA 400076
Email address: `jkv@math.iitb.ac.in`

Contemporary Mathematics
Volume **738**, 2019
https://doi.org/10.1090/conm/738/14878

Decomposition of singular elements of an algebra into product of idempotents, a survey

S. K. Jain and A. Leroy

ABSTRACT. In this paper, we provide a survey of some of the prominent works that were published from the 1960's relating to the problem initially considered by Howie for writing a non-injective mapping as a product of idempotents and by Erdos for writing a singular matrix as a product of idempotents. Besides considering the question of decomposition of nonunits of the algebra of matrices over a special class of rings, the question of a decomposition of a nonunit element of a regular ring into idempotents is included by devoting a full section on the statements of beautiful results of Hannah and O'Meara. The problem of decomposition of nonnegative matrices as a product of nonnegative idempotents is discussed in the last section. Some new results and techniques are also offered. The paper concludes with a list of open questions.

Introduction

This survey article is meant to highlight the important discoveries on the question of presenting non-units of an algebra as a product of idempotents. No attempt is made to give full details, but we direct the interested reader to the references while giving some new proofs.

In the introduction we will go over the history of the problem and give the main discoveries. Open questions are also listed at the end.

This area of research was prompted by the work of Howie ([**35**]) who showed that any non-injective mapping from a finite set into itself is a product of idempotents. J. A. Erdos ([**23**]) considered $n \times n$ matrices over a field and showed that any singular matrix is a product of idempotents. This decomposition into idempotents of matrices or, more generally, of elements of any given Algebra, caught the attention of a number of mathematicians.

Laffey ([**37**], Theorem 1 and 2) proved that this decomposition into idempotents holds for singular matrices over division rings and commutative euclidean domains.

Dawlings (cf. [**14**]) refined the result obtained by Erdos and showed that any singular linear map of an n-dimensional vector space to itself is a product of at most n idempotent linear maps of rank $n-1$.

Key words and phrases. Idempotent, Singular Matrix, Local Ring, Principal Right Ideal Domain, Bézout Domain, Hermite Domain, Injective Module, Regular Ring, Unit Regular Ring, Stable Range 1.

The first author is the corresponding author.

Ballantine ([**11**]) showed that if F is a field and $k, n \in \mathbb{N}$, then a matrix $A \in M_n(F)$ can be written as a product of k idempotent matrices if and only if $rank(I - A) \leq k.nullity(A)$.

Reynolds and Sullivan (cf. [**46**]) studied the question of the minimal number of idempotents needed in the decomposition process for linear transformations of an infinite dimensional vector space.

V. Gould [**16**], [**17**], [**26**] and her coauthors have also considered the problem of decomposition into idempotents using the techniques and machinery of Semigroups. In particular, jointly with Fountain (cf. [**24**]) she studied the case of matrices over principal ideal domains and, using semigroup theory, obtained a series of equivalent conditions for the existence of the decompositions into idempotents. This seminal work was pursued by Ruitenburg (cf [**47**]) who studied the case of matrices over Bézout domains.

Reynolds([**47**]), Sullivan([**49**]), Fountain([**24**]), Ballantine([**11**]) Ruitenburg([**47**]), Dandan-Gould-Quinn ([**17**]), Hannah and O'Meara ([**30**],[**31**]), Alahmadi-Lam-Jain-Leroy([**5**]), among others continued the analysis of the problem.

The results of Ballantine, Reynolds and Sullivan on the number of idempotents needed in a decomposition of an element were improved considerably by Hannah-O'Meara ([**31**], Proposition 1.1) in the following theorem: Let R be a regular ring and let S be the multiplicative subgroup generated by idempotents. Then $\Delta(S) \leq index(R)$, where $\Delta(S)$ denotes the depth of the semigroup S.

As a consequence they showed that in a regular ring, if a is a nilpotent element such that $a^n = 0$, then a is a product of n idempotents.

The contents of our paper are as follows:

In the first section, we consider matrices over arbitrary rings, not necessarily commutative, and prove some new results.

The second section deals with matrices over division rings and valuation rings or more generally, over local rings. We provide new proofs for decomposition of singular matrices over such rings into idempotents.

In the third section we consider quasi-Euclidean domains and give a proof that a singular matrix A (that is $l(A) \neq 0$) is a product of idempotents.

The main results proved by Hannah and O'Meara on von Neumann regular rings whose non units can be decomposed into idempotents are given in Section 4.

The last section 5 deals with the question whether a nonnegative real matrix (like stochastic matrix, (0-1) matrix or totally positive matrix) is a product of nonnegative idempotents.

1. Singular matrices over general rings

In addition to considering rings over which every singular matrix is a product of idempotent matrices, we look at the special types of singular matrices that always admit such a decomposition (whatsoever is the base ring). Such matrices include strictly upper triangular matrices, quasi permutation matrices, or quasi elementary matrices (the latter kind of matrices are considered here for the first time). These matrices can sometimes be used as a tool to prove that other matrices are product of idempotent matrices.

Let us first define two important families of matrices.

DEFINITIONS 1.

(1) *A matrix $A \in M_n(R)$ with coefficients in a ring R will be called a quasi-permutation matrix if each row and each column of A has at most one nonzero element.*

(2) *An elementary matrix is either a permutation matrix or a matrix of the form $I + ae_{ij}$. A quasi elementary matrix is a matrix obtained by replacing at least one row of an elementary matrix by the zero row.*

REMARKS 2.

(a) *A quasi-permutation matrix can be singular and, in this case, it has at least one zero row and one zero column. We will mainly work with rows but the analogous properties for columns also hold (acting on the right with given permutation matrices).*

(b) *Of particular importance in our discussions are the quasi-permutation matrices, denoted by $P_{\sigma,l}$, $\sigma \in S_n$, $l \in \{1, \ldots, n\}$ obtained from P_σ by changing the nonzero element of the lth row of P_σ to 0. We thus have*

$$P_{\sigma,l} = \sum_{\substack{i=1 \\ i \neq l}}^{n} e_{i,\sigma(i)}.$$

(c) *We observe that the $\sigma(l)^{th}$ column of the matrix $P_{\sigma,l}$ is the only column that is zero.*

(d) *It may be of interest to the reader to know the following main steps often used to prove that a matrix is a product of idempotents.*
- *The first step is to check whether left and right annihilators are nonzero.*
- *The second step is to try to show that the given matrix is conjugate to a matrix having a zero row (or column). This always happens if the base ring is a Hermite domain.*
- *Having a zero row, we can assume, using quasi permutation matrices, that the last row (or column) is zero.*
- *Assuming, for instance, that the last row is zero, we try to use an induction on the $(n-1) \times (n-1)$ top left matrix block, either directly or after some more permutations on the columns.*

These steps are used on different occasions in the paper.

(e) *It is also important to notice that sometimes there is a relation between presentations of a noninvertible matrix as a product of idempotent matrices and decompositions of an invertible matrix as a product of elementary matrices. These relations are studied, in particular, by Salce-Zanardo (cf.* [**48**] *) and Facchini-Leroy (cf.* [**20**] *).*

In the next theorem we gather some specimens of singular matrices that are always product of idempotent matrices.

THEOREM 3. *The following singular matrices in $M_n(R)$, $n > 1$, are always product of idempotent matrices, for any ring R.*

(1) *A matrix $A \in M_n(R)$ of the form*

$$\begin{pmatrix} B & C \\ 0 & 0 \end{pmatrix} \text{ or } \begin{pmatrix} B & 0 \\ L & 0 \end{pmatrix}$$

where B is a product of idempotent matrices.

(2) *Any strictly upper (lower) triangular matrix.*

(3) *Any singular* 0-1 *matrix (that is a matrix with* 0, 1 *entries only).*

(4) *For any $\sigma \in S_n$ and any $1 \leq l \leq n$, the matrix $P_{\sigma,l}$ is a product of idempotent matrices.*

(5) *Any diagonal matrix with one zero row.*

(6) *Any quasi permutation matrix is a product of idempotent matrices.*

(7) *Any quasi elementary matrix is a product of idempotent matrices.*

PROOF. If B is a product of idempotent matrices, say $B = E_1 \dots E_l$, then

$$\begin{pmatrix} B & C \\ 0 & 0 \end{pmatrix} = \begin{pmatrix} I_r & C \\ 0 & 0 \end{pmatrix} \begin{pmatrix} E_1 & 0 \\ 0 & I_{n-r} \end{pmatrix} \cdots \begin{pmatrix} E_l & 0 \\ 0 & I_{n-r} \end{pmatrix}$$

Since every matrix on the right hand side is an idempotent matrix we get the desired conclusion.

The second kind of matrix is treated similarly working with columns instead of rows. We may warn the reader that a little care is needed, since our base rings are non commutative (cf. Remark 4).

(2) This is obtained using (1) and an by induction.

(3) If a (0-1)-matrix A is singular this matrix can be written as a product of idempotent matrices in $M_n(\mathbb{Z})$, say $A = E_1 \cdots E_l$ where $E_i^2 = E_i \in M_n(\mathbb{Z})$ since $\mathbb{Z}$ is a commutative Euclidean domain (cf. [**37**]). If φ stands for the usual map $\mathbb{Z} \longrightarrow R$ where $1 \mapsto e$ (e is the identity of R) and $\widehat{\varphi} : M_n(\mathbb{Z}) \longrightarrow M_n(R)$ be its natural extension, we have that $A = \widehat{\varphi}(A) = \widehat{\varphi}(E_1) \cdots \widehat{\varphi}(E_l)$. This gives the result since for any $1 \leq i \leq l$, $\widehat{\varphi}(E_i) \in M_n(R)$ is idempotent.

(4) This is a direct consequence of the Statement 3, since a matrix of the form $P_{\sigma,l}$ is clearly a singular (0-1) matrix.

(5) We proceed by induction on n. If $n = 2$, the diagonal matrix with a zero row can have only two forms as given below. For any $a \in R$:

$$\begin{pmatrix} a & 0 \\ 0 & 0 \end{pmatrix} = \begin{pmatrix} 1 & a \\ 0 & 0 \end{pmatrix} \begin{pmatrix} 0 & 0 \\ 0 & 1 \end{pmatrix} \begin{pmatrix} 1 & 0 \\ 1 & 0 \end{pmatrix} \text{ and}$$

$$\begin{pmatrix} 0 & 0 \\ 0 & a \end{pmatrix} = \begin{pmatrix} 0 & 0 \\ 0 & 1 \end{pmatrix} \begin{pmatrix} 1 & 0 \\ 1 & 0 \end{pmatrix} \begin{pmatrix} 1 & a \\ 0 & 0 \end{pmatrix} \begin{pmatrix} 0 & 0 \\ 0 & 1 \end{pmatrix}$$

We now assume that $n > 2$. Let us first remark that if A is a diagonal matrix with its lth row zero, we have $A = P_{(l,n),l}P_{(l,n)}A$. Since by Statement (4) $P_{(l,n),l}$ is a product of idempotent matrices, it is enough to show that $P_{(l,n)}A$ is a product of idempotent matrices. If $l \neq n$, $P_{(l,n)}A$ is of the form

$$P_{(l,n)}A = \begin{pmatrix} B & C \\ 0 & 0 \end{pmatrix}$$

where B is an $(n-1)\times(n-1)$ diagonal matrix with one row zero (the lth row of B comes from the bottom row of A). In this case the induction hypothesis implies that B is a product of idempotent matrices and Statement (1) above finishes the proof. If $l = n$, the last row and last column of A are zero and $A = AP_{(n-1,n)}P_{(n-1,n),n}$. Since Statement (4) above shows that $P_{(n-1,n),n}$ is a product of idempotent matrices, we only need to show that $AP_{(n-1,n)}$ is a product of idempotent matrices.

This matrix is of the form

$$AP_{(n-1,n)} = \begin{pmatrix} B & C \\ 0 & 0 \end{pmatrix}$$

where B is a diagonal matrix with its last row (and column) zero. The induction hypothesis and Statement (1) give the proof.

(6) For a general quasi permutation matrix, it is enough to remark that such a matrix is a product of the form $diag(a_1, \dots, a_n)P_{\sigma,l}$, where the first matrix stands for a diagonal matrix having some of its entries equal to zero . Note $a_l = 0$. Both matrices of the right hand side are then product of idempotent matrices and hence any quasi-permutation matrix is a product of idempotent matrices.

(7) An $n \times n$ quasi elementary matrix A that is not a quasi permutation matrix must be obtained from a matrix $I_n + ae_{ij}$, $a \in R$, by replacing at least one row by the zero row. As we have seen in the proof of (5) above, we may assume that the zero row is the last one. Then the resulting matrix is itself an idempotent.

□

REMARKS 4.

(1) *One of the very useful techniques for obtaining a decomposition into idempotents is as follows. Let A be a matrix with l^{th} row as zero and let U be an invertible matrix. If U_l stands for the matrix obtained from U by making the lth row as zero, then $A = U_l(U^{-1}A)$. If we know that U_l is a product of idempotent matrices then it is enough that $U^{-1}A$ is a product of idempotents. In particular, since singular quasi-permutation matrices are product of idempotents, when a matrix has a zero row (or column) we can always assume that this is the last row (or column).*

(2) *It is not true, in general, that the transpose of an idempotent matrix is an idempotent. For instance if R is a ring which is not Dedekind finite, we can find $a, b \in R$ such that $ba = 1$ but $ab \neq 1$. We then also have that $(ab)^2 = ab$ but $a^2b \neq a$. It is then easy to check that*

$$A := \begin{pmatrix} ab & a \\ 0 & 0 \end{pmatrix} \quad \textit{and} \quad B = \begin{pmatrix} ab & 0 \\ a & 0 \end{pmatrix}$$

are such that $A^2 = A$ but $B^2 \neq B$. Let us remark that the matrix B is, nevertheless, the product of two idempotent matrices, as the reader may check.

2. Matrices over division rings or local rings

Let us start by mentioning the following results obtained by Erdos for matrices with coefficients in fields and generalized by Laffey for matrices over division rings. We will give a proof of this result which is different and more direct than the original one.

The first part of the following result is a folklore. The second part can be found in [**37**].

LEMMA 5 ([**37**]). *A matrix $A \in M_n(D)$, where D is a division ring, is invertible if and only if $l(A) = 0$ if and only if $r(A) = 0$. If $l(A) \neq 0$, then A is similar to a matrix having its last row equal to zero.*

THEOREM 6 (J.A. Erdos, [**23**], Laffey [**37**], Theorem 1). *Let D be a division ring and $A \in M_n(D)$ be a singular matrix. Then A is a product of idempotent matrices in $M_n(D)$.*

PROOF. By virtue of the above Lemma 5, we may assume that A is of the form:

$$A = \begin{pmatrix} B & C \\ 0 & 0 \end{pmatrix},$$

where B is an $(n-1) \times (n-1)$ matrix. If $n = 2$ and $b = B \neq 0$ then

$$\begin{pmatrix} b & c \\ 0 & 0 \end{pmatrix} = \begin{pmatrix} 1 & b \\ 0 & 0 \end{pmatrix} \begin{pmatrix} 0 & 0 \\ 0 & 1 \end{pmatrix} \begin{pmatrix} 1 & 0 \\ 1 & 0 \end{pmatrix} \begin{pmatrix} 1 & b^{-1}c \\ 0 & 0 \end{pmatrix}.$$

If $b = 0$ then the matrix is strictly upper triangular and the result follows from Theorem 3(2).

Now let us assume that $n > 2$ and that the result is true for singular matrices of size strictly less than n.

- If B is singular then by the induction hypothesis B is a product of idempotent matrices, say $B = E_1 \dots E_r$, for some $r \in \mathbb{N}$ this gives

$$A = \begin{pmatrix} B & C \\ 0 & 0 \end{pmatrix} = \begin{pmatrix} I_{n-1 \times n-1} & C \\ 0 & 0 \end{pmatrix} \begin{pmatrix} E_1 & 0 \\ 0 & 1 \end{pmatrix} \cdots \begin{pmatrix} E_r & 0 \\ 0 & 1 \end{pmatrix}.$$

  This gives the desired conclusion.
- If B is invertible we can write

$$\begin{pmatrix} B & C \\ 0 & 0 \end{pmatrix} = \begin{pmatrix} 0 & B \\ 0 & 0 \end{pmatrix} \begin{pmatrix} 0 & 0 \\ I_{n-1,n-1} & 0 \end{pmatrix} \begin{pmatrix} I_{n-1,n-1} & B^{-1}C \\ 0 & 0 \end{pmatrix}$$

  On the right hand side of this equation, the second matrix is lower triangular, the third matrix is an idempotent, so it remains to show that the first matrix is a product of idempotents.

We just write

$$\begin{pmatrix} 0 & B \\ 0 & 0 \end{pmatrix} = \begin{pmatrix} B' & D \\ 0 & 0 \end{pmatrix}$$

where $B' \in M_{n-1,n-1}(D)$ has its first column zero and D is a column vector. This means that B' is singular and the induction hypothesis implies that B' is in fact a product of idempotents, say $B' = E_1 \dots E_r$ where $E_i^2 = E_i$ for any $1 \leq i \leq r$. We then have

$$\begin{pmatrix} B' & D \\ 0 & 0 \end{pmatrix} = \begin{pmatrix} I_{n-1,n-1} & D \\ 0 & 0 \end{pmatrix} \begin{pmatrix} E_1 & 0 \\ 0 & 1 \end{pmatrix} \cdots \begin{pmatrix} E_r & 0 \\ 0 & 1 \end{pmatrix}.$$

This yields the proof of the theorem. □

In the case of singular matrices, $A \in M_n(k)$ over a field k, a bound on the number of idempotent matrices was given by Ballantine:

THEOREM 7 (Ballantine [**11**]). *Let F be an arbitrary field, n and k be arbitrary positive integers, and A be an $n \times n$ matrix over F. Then A is a product of k idempotent matrices over F if and only if $\rho(I - A) < k\nu(A)$, where ρ and ν denote the rank and the nullity, respectively.*

This was later improved by Hannah and O'Meara (cf. section 4) that includes a larger family of algebras.

Let us also mention that Dawlings (cf. [**14**]) showed that any singular linear map of an n-dimensional vector space to itself is a product of at most n idempotent linear maps of rank $n-1$.

Laffey, in the paper mentioned above ([**37**]), also proved that the decomposition result in Theorem 6 is also true for matrices with coefficients in a Euclidean domain. The case of noncommutative quasi-Euclidean domain was treated by Alahmadi, Jain, Lam and Leroy (cf. the section on quasi-Euclidean domains)

Laffey also showed a kind of converse of the above Theorem 6. Let us mention explicitly his result.

THEOREM 8 ([**38**]). *Let k be a field and A be a finite dimensional k-algebra. Suppose that each non-invertible element of A can be expressed as a product of idempotent elements in A. Then A satisfies one of the following:*
(1) A is isomorphic to $M_n(D)$, the algebra of $n\times n$ matrices over a division algebra D,where D is finite dimensional over F, for some $n\in\mathbb{N}$;
(2) $k=GF(2)$ and A is the direct sum of $s>1$ copies of $GF(2)$;
(3) $k=GF(2)$, $A/J(A)$ is the direct sum $k\oplus k$ where $J(A)$ denotes the Jacobson radical of A and A has a homomorphic image A/I, for some ideal I, isomorphic to the algebra T_2, where T_2 is the upper 2×2 triangular matrices over k.

We will now consider the case of projective-free rings and, in particular, local rings. We will need the following lemma which must be well-known.

LEMMA 9. *A local right Bézout domain is a right valuation domain.*

PROOF. Let $a,b\in R$. We need to prove that either $aR\subseteq bR$ or $bR\subseteq aR$. We know that there exists $g\in R$ such that $aR+bR=gR$. Hence there exist $u,v,a_1,b_1\in R$ such that $au+bv=g, a=ga_1$ and $b=gb_1$. This leads to $g(1-a_1u-b_1v)=0$. Since R is a domain we get that either $g=0$ and then $a=b=0$. If $g\neq 0$ then $1-a_1u-b_1v=0$. Since R is a local ring the second possibility implies that either $a_1u\in U(R)$ or $b_1v\in U(R)$. In the first case we get $aR+bR=gR=ga_1uR=auR\subseteq aR$ and hence $bR\subseteq aR$. Similarly in the second case we obtain $aR\subseteq bR$. □

It was shown in [**2**] that if R is a projective free ring over which matrices with nonzero left and right annihilators are product of idempotents then the ring is actually a domain. In fact R is a Bézout domain.

THEOREM 10. *Let R be a projective-free ring such that any 2×2 matrix A with $l(A)\neq 0$ or $r(A)\neq 0$ is a product of idempotent matrices. Then R is a Bézout domain.*

PROOF. To prove that the ring is a domain, we refer the reader to [**2**] Theorem 9. We will now show that R is right K-Hermite i.e. that for any $a,b\in R$ there exists $Q\in GL_2(R)$ such that $(a,b)Q=(c,0)$. Let $A=\begin{pmatrix} a & b\\ 0 & 0\end{pmatrix}\in\mathbb{M}_2(R)$. Since $l(A)\neq 0$, we can write $A=TE$ for some $T,E\in\mathbb{M}_2(R)$ where $E^2=E\neq I_2$. Then there exists $Q\in \mathrm{GL}_2(R)$ such that $(a,b)=(r,0)\,Q$ for some $r\in R$ (and hence $aR+bR=rR$). In particular, if all matrices $\begin{pmatrix} a & b\\ 0 & 0\end{pmatrix}$ are products of idempotents in $\mathbb{M}_2(R)$, R must be a right K-Hermite ring. A classical result of Amitsur [**6**] shows that this implies that R is a Bézout domain. □

COROLLARY 11. *Let R be a local ring. Suppose that every 2×2 matrix over R having nonzero right or left annihilator is product of idempotents. Then R must be a valuation domain.*

PROOF. Since R is local, R is projective free. The above theorem shows that R is in fact a Bézout domain. Lemma 9 yields the desired conclusion. □

Fountain's results ([**24**]) were based on semigroup techniques. Inspired by Fountain's work Ruitenburg proved that the singular matrices with coefficients in a (not necessarily commutative) Bézout domain are product of idempotent matrices if and only if invertible matrices are product of elementary matrices.

Salce and Zanardo ([**48**]) also considered this relation between the two decompositions. They studied the case of commutative domains but their results were generalized to a noncommutative domains by Facchini and Leroy. To present the latter result we need to introduce a few notions.

DEFINITIONS 12.

- *Let A, B, C be three right R-modules and $\alpha\colon A\to B$, $\beta\colon B\to C$ be homomorphisms. We say that the pair (α,β) is a* consecutive pair *if $im(\alpha)\oplus\ker(\beta)=B$*
- *We say that a ring R is* right n-regular *if for every $n\times n$ invertible matrix $M=(b_{ij})\in M_n(R)$ there exists some $i,j=1,2,\dots,n$ such that $r(b_{ij})=0$.*
- *Let r,n be integers, $0\le r\le n$. For a ring R we define $\mathcal{F}_{n,r}:=\{\,A\subseteq^{\oplus}R^n_R\mid A\cong R^r_R$ and $R^n_R/A\cong R^{n-r}_R\,\}$*

THEOREM 13 (Facchini-Leroy, [**20**]). *Let R be a ring with the IBN property and let $n\ge 1$ be a fixed integer. Suppose that R is m-right regular for every $m=2,3,\dots,n$ and that for any two decompositions of $R^n=A\oplus X=Y\oplus B$ with A,B free right R-modules of ranks, respectively, $n-1,1$, the submodules X,Y are free right R-modules. Then the following conditions are equivalent.*

(S_n) *For every $r=1,2,\dots,n$ and every free direct summands $A\subseteq^{\oplus}R^n_R$ and $B\subseteq^{\oplus}R^n_R$, with A,B free R-modules of rank $r,n-r$ respectively, there exist direct-sum decompositions*

$$R^n_R=A_1\oplus B_1=A_2\oplus B_1=A_2\oplus B_2=\cdots=A_k\oplus B_{k-1}=A_k\oplus B_k$$

with $A=A_1$ and $B=B_k$.

(HI_n) *For every $r=1,2,\dots,n$ and every free direct summands $A\subseteq^{\oplus}R^n_R$ and $B\subseteq^{\oplus}R^n_R$, with A,B free R-modules of rank $r,n-r$ respectively, there exists an endomorphism β of R^n_R with $\mathrm{im}(\beta)=A$ and $\ker(\beta)=B$, such that β is a product $\varepsilon_1\dots\varepsilon_k$ where $\epsilon_1,\dots,\epsilon_k$ are consecutive idempotents $(\mathcal{F}_{n,n-1},\mathcal{F}_{n,1})$-endomorphisms.*

($\mathrm{S}_{n,1}$) *For every free direct summands $A\subseteq^{\oplus}R^n_R$ and $B\subseteq^{\oplus}R^n_R$, of rank $n-1,1$ respectively, there exist direct-sum decompositions*

$$R^n_R=A_1\oplus B_1=A_2\oplus B_1=A_2\oplus B_2=\cdots=A_k\oplus B_{k-1}=A_k\oplus B_k$$

with $A=A_1$ and $B=B_k$.

($\mathrm{HI}_{n,1}$) *For every free direct summands $A \subseteq^{\oplus} R^n_R$ and $B \subseteq^{\oplus} R^n_R$, with A, B free R-modules of rank $n-1, 1$ respectively, there exists an endomorphism β of R^n_R with $\mathrm{im}(\beta) = A$ and $\ker(\beta) = B$, which is a product $\beta = \varepsilon_1 \dots \varepsilon_k$ of consecutive idempotent $(\mathcal{F}_{n,n-1}, \mathcal{F}_{n,1})$-endomorphisms.*

(GE_n) *Every invertible $n \times n$ matrix is a product of elementary matrices.*

Let us remark that the hypotheses of the above theorem are satisfied by a right and left Bézout domain. So that the equivalences of Theorem 13 generalize Ruitenburg results.

Ruitenburg's paper inspired many other works. In particular, some similar results have been proved for quasi-Euclidean rings (cf. [**5**]). Another direction was pursued by Reynolds and Sullivan ([**46**]) who studied the decomposition problem for the linear transformations of an infinite dimensional vector space. Using the ring structure of the algebra of linear transformations Hannah and O'Meara generalized the results to von Neumann regular rings (cf. section 4).

3. Quasi-Euclidean domains

As we mentioned above the case of commutative Euclidean domains was studied by Laffey. This was generalized in [**5**] where noncommutative quasi-Euclidean rings are treated. We recall the results from that paper which are relevant to our discussion and give a complete proof of the decomposition theorem. This proof didn't appear in [**5**]. Firstly, let us recall the definition of a Euclidean pair.

DEFINITION 14. *An ordered pair (a, b) over any ring R is said to be a right Euclidean pair if there exist elements $(q_1, r_1), \dots, (q_{n+1}, r_{n+1}) \in R^2$ (for some $n \geq 0$) such that $a = bq_1 + r_1$, $b = r_1q_2 + r_2$, and*

$$(*) \qquad r_{i-1} = r_i q_{i+1} + r_{i+1} \text{ for } 1 < i \leq n, \text{ with } r_{n+1} = 0.$$

The notion of a left Euclidean pair is defined, similarly. A ring is right quasi-Euclidean if all its pairs are right Euclidean pairs.

We begin this section with the following useful though quite straightforward proposition. This was not explicitly mentioned in [**5**] but will be used in the proof of the Theorem 22.

PROPOSITION 15. *Let R be a right (resp. left) quasi-Euclidean ring. Then*

1. *for any $(a_1, \dots, a_n) \in R^n$, there exist an invertible matrix $Q \in GL_n(R)$ and an element $r \in R$ such that $(a_1, \dots, a_n)Q = (r, 0, \dots, 0)$ (resp.$Q(a_1, \dots, a_n) = (0, \dots, 0, r)$), and*
2. *for any matrix $A \in M_n(R)$ there exists an invertible matrix $Q \in GL_n(R)$ such that AQ (resp. QA) is a lower (resp. upper) triangular matrix.*

PROOF. We will prove the first statement by induction on n. For the case $n = 2$ we remark that, with the notations as in the above definition, we can translate the fact that (a, b) is a Euclidean pair by the following equation

$$(a,b) = (b, r_1)\begin{pmatrix} q_1 & 1 \\ 1 & 0 \end{pmatrix} = (r_1, r_2)\begin{pmatrix} q_2 & 1 \\ 1 & 0 \end{pmatrix}\begin{pmatrix} q_1 & 1 \\ 1 & 0 \end{pmatrix} = \dots = (r_n, 0)P(q_{n+1}) \dots P(q_1)$$

where, for $q \in R$, we have $P(q) = \begin{pmatrix} q & 1 \\ 1 & 0 \end{pmatrix}$. This yields the result.

Let us now consider an element $(a_1,\dots,a_n)\in R^n$ with $n>2$. The case $n=2$ shows that there exist an invertible matrix $Q_1\in GL_2(R)$ and an element $s\in R$ such that $(a_{n-1},a_n)Q_1=(s,0)$. This gives that

$$(a_1,a_2,\dots,a_n)\begin{pmatrix} I_{n-2} & 0\\ 0 & Q_1\end{pmatrix}=(a_1,\dots,a_{n-2},s,0)$$

By induction there exists $Q_2\in GL_{n-1}(R)$ such that $(a_1,\dots,a_{n-2},s)Q_2=(r,0\dots,0)$. This gives

$$(a_1,a_2,\dots,a_n)\begin{pmatrix} I_{n-2} & 0\\ 0 & Q_1\end{pmatrix}\begin{pmatrix} Q_2 & 0\\ 0 & 1\end{pmatrix}=(r,0\dots,0),$$

as desired.

Let $A=(a_{ij})\in M_n(R)$ and, for $1\le i\le n-1$, define $A_i=(a_{i,i},a_{i,i+1},\dots,a_{i,n})$. By Statement (1) we can find $P_i\in GL_{n-i+1}(R)$ and $r_i\in R$ such that $A_iP_i=(r_i,0,\dots,0)$. For $i=1$, we put $Q_1=P_1$ and for $i>1$, we define $Q_i=\begin{pmatrix} I_{i-1} & 0\\ 0 & P_i\end{pmatrix}$. Then we easily obtain that $AQ_1Q_2\cdots Q_{n-1}$ is a lower triangular matrix. □

The next theorem was given in [**5**] and we omit the proof.

THEOREM 16. *Let $(a,b)\in R^2$ be a right Euclidean pair. Then the matrix*

$$\begin{pmatrix} a & b\\ 0 & 0\end{pmatrix}$$

is a product of idempotent matrices.

REMARK 17.

(1) *If the pair (a,b) is left Euclidean instead, a similar decomposition into products of idempotents holds for the matrix $\begin{pmatrix} a & 0\\ b & 0\end{pmatrix}$. We leave the details to the reader.*

(2) *In Theorem 16, the condition that the matrix is a product of idempotents is only necessary but not sufficient for (a,b) to be a Euclidean pair. To see this, let $\theta=\sqrt{-5}$ and $R=\mathbb{Z}[\theta]$ be the full ring of algebraic integers in the number field $\mathbb{Q}[\theta]$. The Dedekind domain R has class number 2, and its class group is generated by the ideal $-2R+(\theta+1)R$ (see [**40**, Example 2.19D]). The matrix $E=\begin{pmatrix} -2 & \theta+1\\ \theta-1 & 3\end{pmatrix}$ over R has trace 1 and determinant 0, and so $E^2=E$. Thus, $A:=\begin{pmatrix} -2 & \theta+1\\ 0 & 0\end{pmatrix}$ has a idempotent factorization, namely, $diag\,(1,0)\,E$. However, the ideal $-2R+(\theta+1)R$ is (by choice) not a principal ideal. In particular, $(-2,\,\theta+1)$ is not a Euclidean pair over R, according to Theorem 6 (a).*

The following theorem appears as Theorem 20 in [**5**].

THEOREM 18. *Let R be a right quasi-Euclidean domain and let $A\in\mathbb{M}_2(R)$ be such that $l.ann\,(A)\neq 0$. Then A is a product of idempotent matrices.*

The following lemmas will be needed in the proof of the next theorem.

LEMMA 19. *If $A\in M_n(R)$ where R is a right Euclidean domain and $l.ann(A)\neq 0$, then A is similar to a matrix with its last row zero.*

LEMMA 20. *Let R be a right quasi-Euclidean ring and suppose that $E \in M_n(R)$ is an idempotent matrix. Then there exists an invertible matrix $P \in M_n(R)$ such that $PEP^{-1} = \begin{pmatrix} I_h & 0 \\ 0 & 0 \end{pmatrix}$, for some $1 \le h \le n$.*

PROOF. Let us first remark that for any $1 \le l < n$ and for any $C \in M_{l\times n}(R)$, we have

$$\begin{pmatrix} I_l & C \\ 0 & 0 \end{pmatrix} = \begin{pmatrix} I_l & -C \\ 0 & I_{n-l} \end{pmatrix} \begin{pmatrix} I_l & 0 \\ 0 & 0 \end{pmatrix} \begin{pmatrix} I_l & C \\ 0 & I_{n-l} \end{pmatrix}.$$

This shows that an idempotent matrix $\begin{pmatrix} I_l & C \\ 0 & 0 \end{pmatrix}$ is always similar to an idempotent of the desired form.

Let us now consider an idempotent matrix Y, $I \neq Y^2 = Y \in M_n(R)$. The above lemma shows that we may assume that Y has its last row zero. Let us write:

$$Y = \begin{pmatrix} B & C \\ 0 & 0 \end{pmatrix} \quad \text{where } B \in M_{n-1,n-1}(R) \text{ and } C \in M_{n-1,1}(R)$$

From $Y^2 = Y$ we get that $B^2 = B$ and $BC = B$. If $B = I_{n-1,n-1}$ it is easy to conclude the result by the above paragraph. if $B \neq I_{n-1,n-1}$, the induction hypothesis shows that there exists an invertible matrix Q such that QBQ^{-1} has the desired form. Thus there exists $l \in \mathbb{N}$ such that:

$$\begin{pmatrix} Q & 0 \\ 0 & 1 \end{pmatrix} \begin{pmatrix} B & C \\ 0 & 0 \end{pmatrix} \begin{pmatrix} Q^{-1} & 0 \\ 0 & 0 \end{pmatrix} = \begin{pmatrix} \begin{pmatrix} I_l & 0 \\ 0 & 0 \end{pmatrix} & QC \\ 0 & 0 \end{pmatrix}.$$

The first paragraph of this proof easily leads to the desired conclusion. □

The following proposition is Proposition 22 in [**5**].

PROPOSITION 21. *Let R be a right quasi-Euclidean domain and $A \in \mathbb{M}_n(R)$. Then $l.ann\,(A) \neq 0$ implies that $r.ann\,(A) \neq 0$.*

We now arrive at the main result of this section which was mentioned in [**5**] without proof.

THEOREM 22. *Let R be a right and left quasi-Euclidean domain. Then every matrix $A \in \mathbb{M}_n(R)$ with $l.ann\,(A) \neq 0$ (equivalently, $r.ann\,(A) \neq 0$) is a product of idempotent matrices.*

PROOF. We proceed by induction on n. If $n = 2$ the result is clear by the Theorem 18. So suppose that $n > 2$ and let $(a_1, \ldots, a_n)A = (0, \ldots, 0)$. Since R is a right Euclidean domain, it is a right Bézout domain and we may assume that $\sum_{i=1}^n a_i R = R$. It follows from Proposition 15(1) that there exists a matrix $Q \in GL_n(R)$ such that its last row is $(a_1, \ldots, a_n)$. By straightforward computations we get that

$$QAQ^{-1} = \begin{pmatrix} B & C \\ 0 & 0 \end{pmatrix} := A'$$

where $B \in M_{n-1}(R)$ and C is a column in R^{n-1} while the last row is the zero row. According to Proposition 15 there exists an invertible matrix Q such that QB is upper triangular. Moreover we have

$$A' = \begin{pmatrix} B & C \\ 0 & 0 \end{pmatrix} = \begin{pmatrix} Q^{-1} & 0 \\ 0 & 0 \end{pmatrix} \begin{pmatrix} QB & QC \\ 0 & 0 \end{pmatrix}.$$

Consider the first matrix on the right hand side. It has a zero column and Remark 4 (1) shows that we can permute the last two columns. An easy induction and Theorem 3 shows that this matrix is a product of idempotent matrices. We just have to deal with the second matrix on the right hand side. This shows that we can assume that the matrix B in A' is in fact upper triangular.
Consider the $(n-1)\times(n-1)$ matrix in the lower right corner of A'. Its bottom row is zero and hence the induction hypothesis implies that this matrix is a product $E_1\dots E_r$ of idempotent matrices $E_i^2=E_i$. Since $E_r^2=E_r$ we have

$$A'=\begin{pmatrix} d_1 & (d_2,\dots,d_n)\\ 0 & E_1\cdots E_r\end{pmatrix}=\begin{pmatrix}1 & 0\\ 0 & E_1\end{pmatrix}\cdots\begin{pmatrix}1 & 0\\ 0 & E_r\end{pmatrix}\begin{pmatrix} d_1 & (d_2,\dots,d_n)\\ 0 & E_r\end{pmatrix}$$

This implies that we only have to show that the last matrix, say M, on the right hand side is a product of idempotent matrices. Lemma 20 shows that there exists an invertible matrix $P\in Gl_{n-1}(R)$ and a natural number $0\le h\le n-1$ such that

$$\begin{pmatrix}1 & 0\\ 0 & P\end{pmatrix}M\begin{pmatrix}1 & 0\\ 0 & P^{-1}\end{pmatrix}=\begin{pmatrix} d_1 & (d_2,\dots,d_n)\\ 0 & PE_rP^{-1}\end{pmatrix}=\begin{pmatrix} d_1 & (d_2,\dots,d_n)\\ 0 & \begin{pmatrix} I_h & 0\\ 0 & 0\end{pmatrix}\end{pmatrix}.$$

It is thus sufficient to consider the case when M is of the form of the last matrix on the right.
We have $1\le h\le n-1$. If $h=n-1$ we would get that E_r is the identity matrix. So we may assume that $0<h<n-1$. Once again if $h<n-2$ then our matrix has the last two rows equal to zero. In this case the induction and Theorem 3 (1) yield the result. So we may assume that $h=n-2$. For $a\in R$ and $2\le i\le n-1$, let us denote $F_i(a)$ the matrix $I_n+ae_{1,i}$ and $\widehat{F_i(a)}$ (resp. $\widehat{I_n}$) the matrix obtained from $F_i(a)$ (resp. I_n) by replacing its last row by a zero row. Now we have, for $2\le i\le n-1$, $M=\widehat{I_n}M=\widehat{F_i(d_i)}F_i(-d_i)M$. Theorem 3 shows that the matrix $\widehat{F_i(d_i)}$ is a product of idempotent matrices and hence we only need to consider matrices of the form $F_i(-d_i)M$, i.e. we may assume that, in the matrix M described above, we have $d_2=d_3=\cdots=d_{n-1}=0$. Now conjugating first the matrix M with $P_{2,n}$ leads to a matrix having its second row zero. This means that our matrix is now of the form:

$$\begin{pmatrix} d_1 & d_n & (0,\dots,0) & 0\\ 0 & 0 & (0,\dots,0) & 0\\ 0 & 0 & X & 0\\ 0 & 0 & 0 & 1\end{pmatrix}$$

(Where X stands for a $((n-3)\times(n-3)$ matrix). Permuting again the second row and the last row we easily conclude, thanks to the induction hypothesis, that the top left $(n-1)\times(n-1)$ corner matrix is a product of idempotents. Theorem 3 (1) then yields the result. □

Let us recall that an IP ring is one over which singular matrices are product of idempotents. Here is a kind of converse of the above Theorem for von Neumann regular rings.

PROPOSITION 23 ([**1**], Theorem 3.2). *An IP ring which is von Neumann regular is unit regular and hence quasi-Euclidean.*

Lezowski ([**37**]) defined Euclidean rings (not necessarily domains) using functions φ from R to $\mathbb{Z}$. Our definition of quasi-Euclidean rings, although similar is independent of the existence of the function φ. He proved that a matrix ring $M_n(R)$ with $n > 1$ over a commutative ring is right Euclidean if and only if the ring R is a principal ideal ring. We may remark that this result is not true for right quasi-Euclidean rings. For example, consider a commutative unit regular ring which is not principal ideal ring. Matrix rings over this ring are quasi-Euclidean. We refer the reader to [**41**] for more details.

4. Hannah and O'Meara

Hannah and O'Meara published several interesting results on the decomposition of nonunit elements of a regular ring into idempotents ([**30**], [**31**], [**42**], [**43**],[**44**]). We will present here some of their results.

THEOREM 24 (Hannah and O'Meara [**30**]). *If an element a of a regular ring R is a product of k idempotents then $(1-a)R \leq k.rann(a)$.*

In the same paper they proved a converse for the unit regular rings:

PROPOSITION 25 ([**30**]). *Let $a \in R$ be an element of a unit regular ring and let k be a positive integer. Then a is a product of k idempotents if and only if $(1-a)R \lesssim k.rann(a)$.*

This leads to the following

COROLLARY 26 ([**30**]). *An element a in a unit regular ring is a product of idempotents if and only if $R.rann(a) = R(1-a)R$.*

In fact it is well-known that in a regular ring a product of unit regular element is unit regular. This was first proved in [**30**] and immediately implies that in a regular rings a product of idempotents is always unit regular. This result is very useful and was used in particular in [**7**].

Hannah and O'Meara also proved the following remarkable result:

THEOREM 27 (Theorem 2.9 in [**30**]). *Let R be one of the following rings: (i) unit regular, (ii) right continuous, or (iii) a factor ring of a right self-injective ring. Then a is a product of idempotents if and only if*

$$R.rann(a) = R(1-a)R = lann(a).R$$

In a more recent paper O'Meara (cf. [**44**]) proved other interesting results related to the decomposition into product of idempotents in a regular ring which is separative. A ring R is separative if for all finitely generated projective modules, A, B

$$A \oplus A \simeq A \oplus B \simeq B \oplus B \quad \text{implies} \quad A \simeq B$$

Equivalently, $nA \cong nB$ for all $n > 1$ implies $A \cong B$

THEOREM 28 (Proposition 2.4, [**44**]). *Let R be a regular ring. Then the separativity of R is equivalent to the fact that an element is a product of idempotents if and only if $R.rann(a) = R(1-a)R = lann(a).R$*

It is worthy to mention that no example of a regular ring that is not separative is known. This is certainly one of the most important open problems in regular rings.

Let us come to another work by Hannah and O'Meara ([**31**]). This work generalizes Reynolds and Sullivan's work and also the Ballantine theorem on the number of idempotents necessary to present a singular element.

For a semigroup S generated by idempotents the depth $\Delta(S)$ of S is the minimum number of idempotents needed to express a general element as a product of idempotents.

THEOREM 29 (Proposition 1.1 and Theorem 1.3 in [**31**]). *Let R be a regular ring and let S be the multiplicative semigroup generated by all its idempotents. Then $\Delta(S) \leq Index(S)$. If R is Dedekind finite, then the equality holds.*

The equality in the above theorem holds when R is right self injective and is not a product $R = R_1 \times R_2$ of a ring R_1 with finite index and a regular ring R_2 of type III. (cf. Goodearl's book [**27**] for the classification of regular rings)

5. Nonnegative Matrices

It is natural to wonder, in the context of real matrices, if a singular nonnegative real matrix is always a product of nonnegative idempotent matrices. We will see that this is not true in general, but there are some positive results.

Firstly let us mention the form of nonnegative idempotent matrices (cf. [**21**]).

THEOREM 30. *Let I be a nonnegative idempotent matrix of rank k. There exists a permutation matrix P such that*

$$PIP^{-1} = \begin{pmatrix} J & JB & 0 & 0 \\ 0 & 0 & 0 & 0 \\ AJ & AJB & 0 & 0 \\ 0 & 0 & 0 & 0 \end{pmatrix} \quad \text{with} \quad J = diag(J_1, \ldots, J_k)$$

where J_i are positive idempotent matrices of rank one (that is, $J_i = x_i y_i^t$ where x_i,y_i are positive vectors and $y_i^t x_i = 1$) and A, B are arbitrary nonnegative matrices of appropriate sizes. Conversely every matrix of the above form is an idempotent of rank k.

This section is heavily based on the recent paper [**4**]. We will only mention the main results.

PROPOSITION 31. *Let $A \in M_n(\mathbb{R}^+)$, $n > 1$, be a nonnegative matrix of rank 1 or 2. Then A is a product of nonnegative idempotent matrices.*

Let us mention that in the case of a matrix of rank 1, it was shown in [**1**] that three idempotent matrices were sufficient.

We now give an example of a nonnegative singular matrix that cannot be written as a product of nonnegative idempotent matrices.

EXAMPLE 32. *Consider the 4×4 matrix*

$$A_\alpha := \begin{pmatrix} \alpha & \alpha & 0 & 0 \\ 0 & 0 & \alpha & \alpha \\ \alpha & 0 & \alpha & 0 \\ 0 & \alpha & 0 & \alpha \end{pmatrix}, \quad \text{where } \alpha \in \mathbb{R}^+,\ \alpha \neq 0.$$

If $A_\alpha = E_1 \cdots E_n$ is a product of nonnegative idempotent nonidentity matrices $E_1, E_2, \ldots, E_n \in M_4(\mathbb{R}^+)$, we get that $A_\alpha = A_\alpha E_n$. For $i = 1, 2, 3, 4$, let us write

(x_i, y_i, z_i, t_i) for the ith row of the matrix E_n. Equating the entries of the rows on both sides we get, after simplification,

For the first row $1 = x_1 + x_2$, $1 = y_1 + y_2$, $0 = z_1 + z_2$ and $0 = t_1 + t_2$.

For the second row $0 = x_3 + x_4$, $0 = y_3 + y_4$, $1 = z_3 + z_4$ and $1 = t_3 + t_4$.

For the third row $1 = x_1 + x_3$, $0 = y_1 + y_3$, $1 = z_1 + z_3$ and $0 = t_1 = t_3$.

For the fourth row $0 = x_2 + x_4$, $1 = y_2 + y_4$, $0 = z_2 + z_4$ and $1 = t_2 + t_4$.

Since all the real numbers x_i, y_i, z_i, t_i must be nonnegative it is easy to conclude from the above equations that the only solution will be $E_n = Id$. This shows that the matrix A_α does not have a presentation as product of nonnegative idempotent matrices. In fact considering $\alpha = 1/2$ we remark that $A_{1/2}$ is a doubly stochastic matrix. Considering the matrix $A_{1/2}A_{1/2}^T$ we may also conclude that even a nonnegative symmetric stochastic matrix cannot always be presented as the product of nonnegative idempotent matrices.

Despite this example there exist special nonnegative matrices which always admit a decomposition into product of nonnegative idempotent matrices. We will conclude this paper giving two such examples. Let us first recall the form of nonnegative matrices having nonnegative von Neumann inverse.

PROPOSITION 33 (cf. [**36**], Theorem 1 and Lemma 2). *If a nonnegative square matrix A admits a nonnegative von Neumann inverse X (i.e. $A = AXA$), then there exists a permutation matrix P such that PAP^T is of the form*

$$PAP^T = \begin{bmatrix} J & JD & 0 & 0 \\ 0 & 0 & 0 & 0 \\ CJ & CJD & 0 & 0 \\ 0 & 0 & 0 & 0 \end{bmatrix},$$

where C and D are nonnegative matrices of suitable sizes and J is a direct sum of matrices of the following three types:

(1) βxy^T where x, y are positive vectors and β is a positive real number.

(2)

$$\begin{bmatrix} 0 & \beta_{12}x_1y_2^T & 0 & 0 & \cdots & 0 \\ 0 & & \beta_{23}x_2y_3^T & 0 & \cdots & 0 \\ \vdots & & & \ddots & \cdots & \vdots \\ 0 & & & 0 & \cdots & \beta_{d-1,d}x_{d-1}y_d^T \\ \beta_{d1}x_dy_1^T & & & 0 & \cdots & 0 \end{bmatrix}$$

where for $1 \le i \le d$ the vectors x_i, y_j are positive and β_{ij} is a positive real number.

(3) the zero matrix.

THEOREM 34. *Let $A \in M_n(\mathbb{R})$, $n \ge 2$, be a singular nonnegative matrix having a nonnegative von Neumann inverse X (i.e. $A = AXA$). Then A is a product of nonnegative idempotent matrices.*

We have seen in Theorem 3 that singular 0-1 matrices are always products of idempotent matrices. In fact, looking at such a matrix as a real matrix one can show much more:

PROPOSITION 35. *Let $A \in M_n(\mathbb{R})$ be a singular nonnegative definite 0-1 matrix. Then A is a product of nonnegative idempotent matrices.*

Let us end this paper with a few questions.

Questions

(1) Is Theorem 22 true for $n = 1$, that is, is it true that singular elements of an indecomposable Quasi-Euclidean ring are product of idempotents?
(2) Is it true that a singular totally nonnegative matrix is a product of nonnegative idempotents? By a result of Cryer [**13**] and also by Goodearl-Lenagan [**28**], it is sufficient to consider a singular totally nonnegative upper or lower triangular matrix.
(3) When can we say that the matrix

$$A = \begin{pmatrix} B & C \\ 0 & 0 \end{pmatrix}$$

where B is invertible is a product of idempotent matrices.
(4) Find classes of nonnegative singular matrices that are always product of nonnegative idempotents.
(5) Does there exists an example of a regular ring R and an element $a \in R$ which is not the product of idempotents and satisfies the condition: $Rr(a) = l(a)R = R(1-a)R$. An answer to this question would in fact solve one of the most fundamental problem in regular rings: does there exist regular non separative ring? (cf. p.129 of [**7**]).
(6) Let R be a ring, $U \in M_n(R)$ an invertible matrix and E, F idempotent matrices not equal to the identity. Is it true that the matrix EUF is a product of idempotent matrices.
(7) What is the structure of an artinian ring R such that every nonunit is a product of idempotents?

References

[1] A. Alahmadi, S. K. Jain, A. Leroy, and A. Sathaye, *Decompositions into products of idempotents*, Electron. J. Linear Algebra **29** (2015), 74–88, DOI 10.13001/1081-3810.2948. MR3414586

[2] A. Alahmadi, S. K. Jain, and A. Leroy, *Decomposition of singular matrices into idempotents*, Linear Multilinear Algebra **62** (2014), no. 1, 13–27, DOI 10.1080/03081087.2012.754439. MR3175397

[3] A. Alahmadi, S. K. Jain, and A. Leroy, *Quasi-permutation singular matrices are products of idempotents*, Linear Algebra Appl. **496** (2016), 487–495, DOI 10.1016/j.laa.2016.01.046. MR3464085

[4] A. Alahmadi, S. K. Jain, and A. Leroy, *When are nonnegative matrices product of nonnegative idempotent matrices?*, Linear Multilinear Algebra **66** (2018), no. 4, 748–758, DOI 10.1080/03081087.2017.1320966. MR3779148

[5] A. Alahmadi, S. K. Jain, T. Y. Lam, and A. Leroy, *Euclidean pairs and quasi-Euclidean rings*, J. Algebra **406** (2014), 154–170, DOI 10.1016/j.jalgebra.2014.02.009. MR3188333

[6] S. A. Amitsur, *Remarks on principal ideal rings*, Osaka Math. J. **15** (1963), 59–69. MR0151484

[7] P. Ara, K. R. Goodearl, K. C. O'Meara, and E. Pardo, *Separative cancellation for projective modules over exchange rings*, Israel J. Math. **105** (1998), 105–137, DOI 10.1007/BF02780325. MR1639739

[8] R. B. Bapat and T. E. S. Raghavan, *Nonnegative matrices and applications*, Encyclopedia of Mathematics and its Applications, vol. 64, Cambridge University Press, Cambridge, 1997. MR1449393

[9] A. Ben-Israel and T. N. E. Greville, *Generalized inverses: Theory and applications*, 2nd ed., CMS Books in Mathematics/Ouvrages de Mathématiques de la SMC, vol. 15, Springer-Verlag, New York, 2003. MR1987382

[10] A. Berman and R. J. Plemmons, *Nonnegative matrices in the mathematical sciences*, Classics in Applied Mathematics, vol. 9, Society for Industrial and Applied Mathematics (SIAM), Philadelphia, PA, 1994. Revised reprint of the 1979 original. MR1298430

[11] C. S. Ballantine, *Products of idempotent matrices*, Linear Algebra and Appl. **19** (1978), no. 1, 81–86, DOI 10.1016/0024-3795(78)90006-x. MR0472862

[12] B. Brown and N. H. McCoy, *The maximal regular ideal of a ring*, Proc. Amer. Math. Soc. **1** (1950), 165–171, DOI 10.2307/2031919. MR0034757

[13] C. W. Cryer, *The LU-factorization of totally positive matrices*, Linear Algebra and Appl. **7** (1973), 83–92, DOI 10.1016/0024-3795(73)90039-6. MR0311688

[14] R. J. H. Dawlings, *Products of idempotents in the semigroup of singular endomorphisms of a finite-dimensional vector space*, Proc. Roy. Soc. Edinburgh Sect. A **91** (1981/82), no. 1-2, 123–133, DOI 10.1017/S0308210500012683. MR648922

[15] R. J. H. Dawlings, *The idempotent generated subsemigroup of the semigroup of continuous endomorphisms of a separable Hilbert space*, Proc. Roy. Soc. Edinburgh Sect. A **94** (1983), no. 3-4, 351–360, DOI 10.1017/S0308210500015717. MR709728

[16] Y. Dandan and V. Gould, *Free idempotent generated semigroups and endomorphism monoids of independence algebras*, Semigroup Forum **93** (2016), no. 3, 535–553, DOI 10.1007/s00233-016-9802-0. MR3572416

[17] Y. Dandan, V. Gould, and T. Quinn-Gregson, *Free idempotent generated semigroups: subsemigroups, retracts and maximal subgroups*, Comm. Algebra **46** (2018), no. 5, 2264–2277, DOI 10.1080/00927872.2017.1378893. MR3799208

[18] D. Yang, I. Dolinka, and V. Gould, *Free idempotent generated semigroups and endomorphism monoids of free G-acts*, J. Algebra **429** (2015), 133–176, DOI 10.1016/j.jalgebra.2014.12.041. MR3320618

[19] D. Ž. Djoković, *Note on a theorem on singular matrices*, Canad. Math. Bull. **11** (1968), 283–284, DOI 10.4153/CMB-1968-034-9. MR0230742

[20] A. Facchini and A. Leroy, *Elementary matrices and products of idempotents*, Linear Multilinear Algebra **64** (2016), no. 10, 1916–1935, DOI 10.1080/03081087.2015.1127885. MR3521148

[21] P. Flor, *On groups of non-negative matrices*, Compositio Math. **21** (1969), 376–382. MR0257115

[22] J. Fountain and A. Lewin, *Products of idempotent endomorphisms of an independence algebra of finite rank*, Proc. Edinburgh Math. Soc. (2) **35** (1992), no. 3, 493–500, DOI 10.1017/S0013091500005769. MR1187010

[23] J. A. Erdos, *On products of idempotent matrices*, Glasgow Math. J. **8** (1967), 118–122, DOI 10.1017/S0017089500000173. MR0220751

[24] J. Fountain, *Products of idempotent integer matrices*, Math. Proc. Cambridge Philos. Soc. **110** (1991), no. 3, 431–441, DOI 10.1017/S030500410007050X. MR1120478

[25] J. Fountain and A. Lewin, *Products of idempotent endomorphisms of an independence algebra of infinite rank*, Math. Proc. Cambridge Philos. Soc. **114** (1993), no. 2, 303–319, DOI 10.1017/S0305004100071607. MR1230134

[26] J. Fountain and V. Gould, *Products of idempotent endomorphisms of relatively free algebras with weak exchange properties*, Proc. Edinb. Math. Soc. (2) **50** (2007), no. 2, 343–362, DOI 10.1017/S0013091504000902. MR2334951

[27] K. R. Goodearl, *von Neumann regular rings*, Monographs and Studies in Mathematics, vol. 4, Pitman (Advanced Publishing Program), Boston, Mass.-London, 1979. MR533669

[28] K. R. Goodearl and T. H. Lenagan, *LU decomposition of totally nonnegative matrices*, Linear Algebra Appl. **436** (2012), no. 7, 2554–2566, DOI 10.1016/j.laa.2011.11.005. MR2890014

[29] N. Jacobson, *Structure of rings*, American Mathematical Society Colloquium Publications, Vol. 37. Revised edition, American Mathematical Society, Providence, R.I., 1964. MR0222106

[30] J. Hannah and K. C. O'Meara, *Products of idempotents in regular rings. II*, J. Algebra **123** (1989), no. 1, 223–239, DOI 10.1016/0021-8693(89)90044-6. MR1000485

[31] J. Hannah and K. C. O'Meara, *Depth of idempotent-generated subsemigroups of a regular ring*, Proc. London Math. Soc. (3) **59** (1989), no. 3, 464–482, DOI 10.1112/plms/s3-59.3.464. MR1014867

[32] J. M. Howie, *An introduction to semigroup theory*, L.M.S. Monographs, No. 7, Academic Press [Harcourt Brace Jovanovich, Publishers], London-New York, 1976. MR0466355

[33] J. M. Howie, *Some subsemigroups of infinite full transformation semigroups*, Proc. Roy. Soc. Edinburgh Sect. A **88** (1981), no. 1-2, 159–167, DOI 10.1017/S0308210500017364. MR611308

[34] J. M. Howie, *Products of idempotents in finite full transformation semigroups*, Proc. Roy. Soc. Edinburgh Sect. A **86** (1980), no. 3-4, 243–254, DOI 10.1017/S0308210500012154. MR592552

[35] J. M. Howie, *The subsemigroup generated by the idempotents of a full transformation semigroup*, J. London Math. Soc. **41** (1966), 707–716, DOI 10.1112/jlms/s1-41.1.707. MR0219649

[36] S. K. Jain and V. K. Goel, *Nonnegative matrices having nonnegative Drazin pseudoinverses*, Linear Algebra Appl. **29** (1980), 173–183, DOI 10.1016/0024-3795(80)90238-4. MR562757

[37] T. J. Laffey, *Products of idempotent matrices*, Linear and Multilinear Algebra **14** (1983), no. 4, 309–314, DOI 10.1080/03081088308817567. MR724380

[38] T. J. Laffey, *A characterisation of full matrix algebras*, Linear and Multilinear Algebra **15** (1984), no. 2, 105–108, DOI 10.1080/03081088408817581. MR740664

[39] T. J. Laffey, *A note on embedding finite semigroups in finite semibands*, Quart. J. Math. Oxford Ser. (2) **34** (1983), no. 136, 453–454, DOI 10.1093/qmath/34.4.453. MR723281

[40] T. Y. Lam, *Lectures on Modules and Rings*, Graduate Texts in Mathematics, vol. 189, Springer-Verlag, New York, 1999. MR1653294

[41] P. Lezowski, *On some Euclidean properties of matrix algebras*, J. Algebra **486** (2017), 157–203, DOI 10.1016/j.jalgebra.2017.05.018. MR3666211

[42] K. C. O'Meara, *Products of idempotents in regular rings*, Glasgow Math. J. **28** (1986), no. 2, 143–152, DOI 10.1017/S0017089500006467. MR848421

[43] K. C. O'Meara, *Simple regular rings satisfying weak comparability*, J. Algebra **141** (1991), no. 1, 162–186, DOI 10.1016/0021-8693(91)90210-Y. MR1118322

[44] K. C. O'Meara, *Products of idempotents in separative regular rings*, J. Algebra Appl. **13** (2014), no. 8, 1450071, 14 pp., DOI 10.1142/S0219498814500716. MR3225138

[45] K. P. S. Bhaskara Rao, *Products of idempotent matrices over integral domains*, Linear Algebra Appl. **430** (2009), no. 10, 2690–2695, DOI 10.1016/j.laa.2008.11.018. MR2509850

[46] M. A. Reynolds and R. P. Sullivan, *Products of idempotent linear transformations*, Proc. Roy. Soc. Edinburgh Sect. A **100** (1985), no. 1-2, 123–138, DOI 10.1017/S0308210500013688. MR801849

[47] W. Ruitenburg, *Products of idempotent matrices over Hermite domains*, Semigroup Forum **46** (1993), no. 3, 371–378, DOI 10.1007/BF02573579. MR1206214

[48] L. Salce and P. Zanardo, *Products of elementary and idempotent matrices over integral domains*, Linear Algebra Appl. **452** (2014), 130–152, DOI 10.1016/j.laa.2014.03.042. MR3201094

[49] E. Seneta, *Nonnegative matrices and Markov chains*, 2nd ed., Springer Series in Statistics, Springer-Verlag, New York, 1981. MR719544

Department of Mathematics, King Andulaziz University Jeddah, SA, and Ohio University, USA
Email address: jain@ohio.edu

Department of Mathematics, University of Artois. Lens, France
Email address: a.leroy55@gmail.com

Contemporary Mathematics
Volume **738**, 2019
https://doi.org/10.1090/conm/738/14879

Some results on integrally closed domains

Anuj Jakhar, Sudesh K. Khanduja, and Neeraj Sangwan

ABSTRACT. Let K_1, K_2 be finite separable extensions of a field K which are linearly disjoint over K. In 2017, we proved that if the integral closures S_1, S_2 of a valuation ring R_v of K with perfect residue field, in K_1, K_2 respectively are free R_v-modules and the composite ring S_1S_2 is integrally closed, then the discriminant of either S_1/R_v or S_2/R_v is the unit ideal (cf. [J. Pure Appl. Algebra **222** (2018), 3560–3565]). From this result it was deduced that in particular when K_1, K_2 are algebraic number fields with the composite ring $A_{K_1}A_{K_2}$ integrally closed, A_{K_i} being the ring of algebraic integers of K_i, then the relative discriminants of K_1/K and K_2/K are coprime. In this paper, we show that the converse of both the above results holds.

1. Introduction

Let R be an integrally closed domain and θ be an element of an integral domain containing R with θ integral over R having minimal polynomial $F(x)$ over the quotient field of R. Kummer proved that if $R[\theta]$ is an integrally closed domain, then the maximal ideals of $R[\theta]$ which lie over a maximal ideal $\wp$ of R are in one-to-one correspondence with the monic irreducible factors over $R/\wp$ of the polynomial $\overline{F}(x)$ obtained on replacing the coefficients of $F(x)$ modulo $\wp$; indeed if $g(x)$ belonging to $R[x]$ is irreducible modulo $\wp$ which divides $F(x)$ modulo $\wp$, then the corresponding maximal ideal of $R[\theta]$ is the one generated by $\wp$, $g(\theta)$ (see [**15**, Chapter V, Theorem 34]).

History of the problem. The question "when is $R[\theta]$ integrally closed" has inspired many mathematicians (cf. [**1**], [**6**], [**9**], [**14**]). This problem is closely related with the existence of a power basis of an algebraic number field. Recall that a power basis of an algebraic number field K is a $\mathbb{Z}$-basis of the ring of algebraic integers of K consisting of powers of a single element; indeed α would be such an element if and only if $\mathbb{Z}[\alpha]$ is integrally closed in its quotient field K. If A_K denotes the ring of algebraic integers of an algebraic number field $K = \mathbb{Q}(\theta)$ with θ an algebraic integer and $\mathbb{Z}_{(p)}$ denotes the localization of $\mathbb{Z}$ at a nonzero prime ideal $p\mathbb{Z}$, then it is immediate from Lagrange's theorem for finite groups that when a prime p does not divide the index of the subgroup $\mathbb{Z}[\theta]$ in A_K, then $A_K \subseteq \mathbb{Z}_{(p)}[\theta]$. The converse assertion also holds because if p divides $[A_K : \mathbb{Z}[\theta]]$, then by Cauchy's

2010 *Mathematics Subject Classification.* Primary : 12J10; 11R04.

The second author is the corresponding author.

The financial support from the respective institutions is gratefully acknowledged by the authors. The second author is thankful to Indian National Science Academy for the fellowship.

theorem, the group $A_K/\mathbb{Z}[\theta]$ has an element $\xi + \mathbb{Z}[\theta]$ of order p, in which case the element ξ of A_K does not belong to $\mathbb{Z}_{(p)}[\theta]$. Thus p does not divide $[A_K : \mathbb{Z}[\theta]]$ if and only if $A_K \subseteq \mathbb{Z}_{(p)}[\theta]$, which is the same as requiring that $\mathbb{Z}_{(p)}[\theta]$ is an integrally closed domain. In 1878, Dedekind gave a necessary and sufficient criterion to be satisfied by the minimal polynomial $F(x)$ of θ over $\mathbb{Q}$ so that $p \nmid [A_K : \mathbb{Z}[\theta]]$. He proved that if $\overline{F}(x) = \bar{g}_1(x)^{e_1} \cdots \bar{g}_t(x)^{e_t}$ is the factorization of the polynomial $\overline{F}(x)$ obtained by replacing coefficients of $F(x)$ modulo p as a product of powers of distinct irreducible polynomials over $\mathbb{Z}/p\mathbb{Z}$ with each $g_i(x)$ belonging to $\mathbb{Z}[x]$ monic, then $\mathbb{Z}_{(p)}[\theta]$ is integrally closed if and only if for each i, either $e_i = 1$ or $\bar{g}_i(x) \nmid \overline{M}(x)$, where $M(x) = \frac{1}{p}(F(x) - \prod_{i=1}^{t} g_i(x)^{e_i})$ (see [**2**, Theorem 6.1.4], [**3**]). Since $\mathbb{Z}_{(p)}$ is the valuation ring of the p-adic valuation of rationals, the above criterion provided a motivation to investigate the problem "when is a simple ring extension $R_v[\theta]$ of the valuation ring R_v of a Krull valuation v integrally closed" (see the next section for definitions of Krull valuation, valuation ring and residue field). The above problem was first solved for discrete valuation rings (see [**10**, Theorem 1.1]) and then for arbitrary valuation rings by Ershov in 2006 (cf. [**6**], [**11**]). Indeed he proved the following result in this direction.

Theorem (Generalized Dedekind Criterion). *Let v be a Krull valuation of a field with valuation ring R_v having maximal ideal M_v. For $g(x) \in R_v[x]$, let $\bar{g}(x)$ denote the polynomial obtained on replacing each coefficient of $g(x)$ by its image under the canonical homomorphism from R_v onto R_v/M_v. Let $F(x) \in R_v[x]$ be a monic irreducible polynomial having a root θ in its splitting field and $\overline{F}(x) = \bar{g}_1(x)^{e_1} \cdots \bar{g}_t(x)^{e_t}$ be the factorization of $\overline{F}(x)$ into a product of powers of distinct irreducible polynomials over R_v/M_v with $g_i(x) \in R_v[x]$ monic. Then $R_v[\theta]$ is integrally closed if and only if either $e_i = 1$ for each i or some $e_j > 1$, in which case M_v is a principal ideal say generated by π and $\bar{g}_j(x)$ does not divide $\overline{M}(x)$ for such an index j, where $M(x) = \frac{1}{\pi}(F(x) - g_1(x)^{e_1} \cdots g_t(x)^{e_t})$.*

In 2015, the following equivalent version of the above criterion was proved (see [**9**]).

Theorem 1.A. *Let $v, R_v, M_v, \theta, F(x), g_i(x)$ be as in the above theorem. Then $R_v[\theta]$ is integrally closed if and only if $F(x)$ does not belong to the ideal $\langle M_v, g_i(x)\rangle^2$ in $R_v[x]$ for any i, $1 \leq i \leq t$.*

With R_v as the valuation ring of a Krull valuation v and θ a root of an irreducible trinomial $x^n + ax^m + b$ belonging to $R_v[x]$, using Generalized Dedekind Criterion, we gave explicit necessary and sufficient conditions involving only a, b, m, n for $R_v[\theta]$ to be integrally closed (see [**7**, Theorem 1.1]). As an application we proved in [**7**] the following Theorem 1.B providing a partial converse of the well known result which states that if K_1, K_2 are algebraic number fields with coprime discriminants, then K_1, K_2 are linearly disjoint over the field $\mathbb{Q}$ of rational numbers and the composite ring $A_{K_1}A_{K_2}$ is integrally closed, A_{K_i} being the ring of algebraic integers of K_i (cf. [**12**, Theorem 4.26]).

Theorem 1.B. *Let K_1, K_2 be algebraic number fields with K_2 a quadratic field not contained in K_1. If the composite ring $A_{K_1}A_{K_2}$ is integrally closed, then the discriminants of K_1 and K_2 are coprime.*

Motivation of the problem. If K_1, K_2 are as in the above theorem, then K_1, K_2 are linearly disjoint over the field $\mathbb{Q}$ of rational numbers. Recall that if L_1, L_2 are field extensions of a field K, both embedded in some field extension Ω of K, then they are said to be linearly disjoint over K in Ω if any K-basis of L_1 is linearly independent over L_2; when this is so, $L_1 \cap L_2 = K$.

Theorem 1.B led us to consider the following problem.

If K_1, K_2 are algebraic number fields linearly disjoint over $\mathbb{Q}$ for which $A_{K_1}A_{K_2}$ is integrally closed, then is it true that the discriminants of K_1 and K_2 are coprime?

In 2017, we proved that the answer to the above question is in the affirmative and the following result was obtained in this direction (see [**8**, Corollary 1.2]).

Theorem 1.C. *Let K_1, K_2 be algebraic number fields which are linearly disjoint over $K = K_1 \cap K_2$ such that the composite ring $A_{K_1}A_{K_2}$ is integrally closed. Then the relative discriminants*[1] *of the extensions K_1/K and K_2/K are coprime.*

In fact the above theorem was deduced as a corollary of a more general result proved in [**8**, Theorem 1.1] which will be stated after introducing a general notion of discriminant.

Definition. Let R be an integrally closed domain with quotient field K and S be the integral closure of R in a finite separable extension L of K of degree n. Assume that S is a free R-module of rank n. As usual the discriminant of S/R to be denoted by $d(S/R)$ is defined to be the ideal in R generated by the determinant of the $n \times n$ matrix $(Tr_{L/K}(\beta_i\beta_j))_{ij}$, where $\{\beta_1, \cdots, \beta_n\}$ is an R-basis of S and Tr is the trace map. As in [**12**, Proposition 2.9(ii)], it can be easily seen that for any other R-basis $\{\beta_1', \beta_2', \cdots \beta_n'\}$ of S, the determinants of the matrices $(Tr_{L/K}(\beta_i\beta_j))_{ij}$ and $(Tr_{L/K}(\beta_i'\beta_j'))_{ij}$ differ multiplicatively by a unit of R. So $d(S/R)$ is well defined.

With the above notation, the main result of [**8**] can be stated as follows.

Theorem 1.D. *Let v be a Krull valuation of a field K with perfect residue field and K_1, K_2 be finite separable extensions of K which are linearly disjoint over K. Let S_1, S_2 denote the integral closures of the valuation ring R_v of v in K_1, K_2 respectively. If S_1, S_2 are free R_v-modules and the composite ring S_1S_2 is integrally closed, then either $d(S_1/R_v)$ or $d(S_2/R_v)$ is the unit ideal.*

Statement of Results. In this paper, our aim is to show that the converse of Theorem 1.C as well as of Theorem 1.D holds. Precisely stated, we prove the following results.

[1]As in [**12**], the relative discriminant of an extension L/K of algebraic number fields is the norm relative to L/K of the inverse of the fractional ideal $\{\lambda \in L \mid Tr_{L/K}(\lambda A_L) \subseteq A_K\}$ of the ring A_L of algebraic integers of L.

THEOREM 1.1. *Let v be a Krull valuation of a field K and K_1, K_2 be finite separable extensions of K which are linearly disjoint over K. Let S_1, S_2, S denote the integral closures of the valuation ring R_v of v in K_1, K_2, K_1K_2 respectively. If S_1, S_2 are free R_v-modules, then $d(S_i/R_v)S \subseteq S_1S_2$ for $i = 1, 2$. In particular if one of $d(S_1/R_v)$ or $d(S_2/R_v)$ is the unit ideal, then S_1S_2 is integrally closed.*

As an application of the above theorem, we prove the following theorem. Indeed we prove a slightly more general result stated in Theorem 3.1.

THEOREM 1.2. *Let K_1, K_2 be algebraic number fields which are linearly disjoint over $K = K_1 \cap K_2$ such that the relative discriminants of the extensions K_1/K and K_2/K are coprime. Then the composite ring $A_{K_1}A_{K_2}$ is integrally closed.*

2. Some Definitions.

By a Krull valuation v of a field K, we mean a mapping v from K onto $G \cup \{\infty\}$, where G is a totally ordered additive abelian group, such that for all a, b in K, the following properties are satisfied:
(i) $v(a) = \infty$ if and only if $a = 0$;
(ii) $v(ab) = v(a) + v(b)$;
(iii) $v(a + b) \geq \min\{v(a), v(b)\}$.

The subset

$$R_v = \{a \in K \mid v(a) \geq 0\}$$

is a subring of K, called the valuation ring of v and usually denoted by R_v. It is a local ring with unique maximal ideal

$$M_v = \{a \in K \mid v(a) > 0\}$$

and R_v/M_v is called the residue field of v.

As in [**5**, Theorem 3.1.3], it can be easily seen that the valuation ring R_v of v is integrally closed.

Let A be a Dedekind domain with quotient field K and L be finite separable extension of K. Let B be the integral closure of A in L. As in [**13**, Chapter 3], the relative discriminant of L/K (with respect to A) is the norm relative to L/K of the inverse of the fractional ideal B^* defined by $B^* = \{\lambda \in L \mid Tr_{L/K}(\lambda B) \subseteq A\}$.

3. Proof of Theorems 1.1, 1.2.

Proof of Theorem 1.1. For the first assertion of the theorem, it is enough to prove that $d(S_1/R_v)S \subseteq S_1S_2$ in view of symmetry. Let $\{\alpha_1, \alpha_2, \cdots, \alpha_m\}$ be an R_v-basis of S_1 and $\{\beta_1, \beta_2, \cdots, \beta_n\}$ be an R_v-basis of S_2. Since $[K_1K_2 : K] = mn$, the family $\{\alpha_j\beta_k \mid 1 \leq j \leq m, 1 \leq k \leq n\}$ is a K-vector space basis of K_1K_2. Let ω be any element of S. We can write

$$\omega = \sum_{j,k} \frac{r_{jk}}{r}\alpha_j\beta_k, \ r_{jk}, r \in R_v. \tag{3.1}$$

Since $[K_1K_2 : K] = mn$, every K-embedding σ of K_1 into its algebraic closure can be extended to K_1K_2 acting trivially on K_2 and hence

$$\sigma(\omega) = \sum_{j,k} \frac{r_{jk}}{r}\sigma(\alpha_j)\beta_k. \tag{3.2}$$

Set $x_j = \sum_{k=1}^{n} \frac{r_{jk}}{r}\beta_k$. On applying to (3.1) the K_2-isomorphisms $\sigma_1, \sigma_2, \cdots, \sigma_m$ of K_1K_2 into the algebraic closure of K and keeping in mind (3.2), we obtain m equations

$$\sum_{j=1}^{m} \sigma_i(\alpha_j)x_j = \sigma_i(\omega), 1 \leq i \leq m.$$

On solving for x_j by Cramer's rule, we see that $x_j = \frac{\gamma_j}{\delta}$ with $\delta = \det(A)$, where A is the matrix having $(i,j)^{th}$ entry $\sigma_i(\alpha_j)$ and γ_j is the determinant of the matrix obtained on replacing the entries of the jth column of A by $\sigma_1(\omega), \cdots, \sigma_m(\omega)$ respectively. So δ, γ_j and $\delta\gamma_j$ are integral over R_v for $1 \leq j \leq m$. Now δ^2 being the determinant of the matrix $A^tA = (Tr_{K_1/K}(\alpha_i\alpha_j)_{ij})$ belongs to R_v. Consequently $\delta\gamma_j = \delta^2 x_j = \sum_{k=1}^{n} \frac{\delta^2 r_{jk}}{r}\beta_k$ belongs to S_2. Since $\{\beta_1, \cdots, \beta_n\}$ is an R_v-basis of S_2, we conclude that $\delta^2\frac{r_{jk}}{r} \in R_v$ for each pair j,k. Therefore $\delta^2\omega = \sum_j\sum_k \frac{\delta^2 r_{jk}}{r}\beta_k\alpha_j \in S_2S_1$. As ω is an arbitrary element of S and δ^2 generates the ideal $d(S_1/R_v)$, we conclude that $d(S_1/R_v)S \subseteq S_1S_2$. This proves the first assertion of the theorem. The second follows immediately from the first.

Proof of Theorem 1.2. Let $D(K_i/K)$ stand for the relative discriminant of K_i/K. Suppose to the contrary that $A_{K_1}A_{K_2}$ is not integrally closed. Let z be an element of $A_{K_1K_2} \setminus A_{K_1}A_{K_2}$. Consider the proper ideal

$$I = \{r \in A_K \mid rz \in A_{K_1}A_{K_2}\} \tag{3.3}$$

of A_K. Then $I \subseteq \wp$ for some maximal ideal $\wp$ of A_K. Let v stand for the $\wp$-adic valuation of K defined for any non-zero $\alpha \in A_K$ to be the highest power of $\wp$ dividing the ideal αA_K. Let $S_1^{(\wp)}$ and $S_2^{(\wp)}$ denote the integral closures in K_1, K_2 respectively of the valuation ring R_v of v which is the localization of A_K at $\wp$. Set $T = A_K \setminus \wp$. It can be easily seen that $S_i^{(\wp)}$ equals the ring of fractions $T^{-1}A_{K_i}$ of A_{K_i} with respect to the multiplicatively closed set T. As R_v is a discrete valuation ring with K_i/K a separable extension, we see that $S_i^{(\wp)}$ is a free R_v-module in view of [**4**, Corollary 18.7] and so $d(S_i^{(\wp)}/R_v)$ is defined. It is well known that $D(K_1/K)R_v = d(S_1^{(\wp)}/R_v)$ and $D(K_2/K)R_v = d(S_2^{(\wp)}/R_v)$ (cf. [**13**, Chapter 3, Proposition 9]). By hypothesis $D(K_1/K)$ and $D(K_2/K)$ are coprime and hence so are the ideals $d(S_1^{(\wp)}/R_v)$, $d(S_2^{(\wp)}/R_v)$ of the discrete valuation ring R_v. Consequently either $d(S_1^{(\wp)}/R_v)$ or $d(S_2^{(\wp)}/R_v)$ is the unit ideal. Applying Theorem 1.1, we see that the composite ring $S_1^{(\wp)}S_2^{(\wp)}$ is integrally closed and hence it contains the integral closure $A_{K_1K_2}$ of A_K in K_1K_2. Keeping in mind that $S_i^{(\wp)} = T^{-1}A_{K_i}$, we can write the element z of $A_{K_1K_2} \subseteq S_1^{(\wp)}S_2^{(\wp)}$ as $z = \sum_j \frac{a_jb_j}{t}$ where $a_j \in A_{K_1}, b_j \in A_{K_2}$ and $t \in T$. Therefore $tz \in A_{K_1}A_{K_2}$; consequently t belongs to the ideal I defined by (3.3). This is impossible as $I \subseteq \wp$ and $t \in T = A_K \setminus \wp$. This contradiction proves that $A_{K_1}A_{K_2} = A_{K_1K_2}$ and hence the theorem.

It may be pointed out that the above proof can be carried over verbatim to prove the following more general result.

THEOREM 3.1. *Let A be a Dedekind domain with quotient field K and K_1, K_2 be finite separable extensions of K which are linearly disjoint over K. Let A_1, A_2 denote the integral closures of A in K_1, K_2 respectively. If the relative discriminants of K_1/K and K_2/K (with respect to A) are coprime, then A_1A_2 is integrally closed.*

EXAMPLE 3.2. Let d be a squarefree integer and $u \neq 1, d$ be an integer dividing d such that $u \equiv 1 \pmod 4$. Then an integral basis of $\mathbb{Q}(\sqrt{u}, \sqrt{d})$ is $\{1, \sqrt{u}, \sqrt{d/u}, \sqrt{d}\}$ in view of Theorem 1.2 as $\mathbb{Q}(\sqrt{u}, \sqrt{d})$ is the compositum of the fields $\mathbb{Q}(\sqrt{u})$, $\mathbb{Q}(\sqrt{d/u})$, which have coprime discriminants.

References

[1] M. E. Charkani and A. Deajim, *Generating a power basis over a Dedekind ring*, J. Number Theory **132** (2012), no. 10, 2267–2276, DOI 10.1016/j.jnt.2012.04.006. MR2944754

[2] H. Cohen, *A course in computational algebraic number theory*, Graduate Texts in Mathematics, vol. 138, Springer-Verlag, Berlin, 1993. MR1228206

[3] R. Dedekind, *Über den Zusammenhang zwischen der Theorie der Ideale und der Theorie der höheren Kongruenzen*, Götttingen Abh. **23** (1878), 1–23.

[4] O. Endler, *Valuation theory*, Springer-Verlag, New York-Heidelberg, 1972. MR0357379

[5] A. J. Engler and A. Prestel, *Valued fields*, Springer Monographs in Mathematics, Springer-Verlag, Berlin, 2005. MR2183496

[6] Y. L. Ershov, *A Dedekind criterion for arbitrary valuation rings*, Doklady Mathematics (2006), 650–652.

[7] A. Jakhar, S. K. Khanduja, and N. Sangwan, *On integrally closed simple extensions of valuation rings*, J. Pure Appl. Algebra **222** (2018), no. 4, 889–899, DOI 10.1016/j.jpaa.2017.05.012. MR3720858

[8] A. Jakhar, S. K. Khanduja, and N. Sangwan, *On the compositum of integral closures of valuation rings*, J. Pure Appl. Algebra **222** (2018), no. 11, 3560–3565, DOI 10.1016/j.jpaa.2017.12.023. MR3806741

[9] S. K. Khanduja and B. Jhorar, *When is $R[\theta]$ integrally closed?*, J. Algebra Appl. **15** (2016), no. 5, 1650091, 7, DOI 10.1142/S0219498816500912. MR3479450

[10] M. Kumar and S. K. Khanduja, *A generalization of Dedekind criterion*, Comm. Algebra **35** (2007), no. 5, 1479–1486, DOI 10.1080/00927870601168897. MR2317622

[11] S. K. Khanduja and M. Kumar, *On Dedekind criterion and simple extensions of valuation rings*, Comm. Algebra **38** (2010), no. 2, 684–696, DOI 10.1080/00927870902829080. MR2598906

[12] W. Narkiewicz, *Elementary and analytic theory of algebraic numbers*, 3rd ed., Springer Monographs in Mathematics, Springer-Verlag, Berlin, 2004. MR2078267

[13] J.-P. Serre, *Local fields*, Graduate Texts in Mathematics, vol. 67, Springer-Verlag, New York-Berlin, 1979. Translated from the French by Marvin Jay Greenberg. MR554237

[14] K. Uchida, *When is $Z[\alpha]$ the ring of the integers?*, Osaka J. Math. **14** (1977), no. 1, 155–157. MR0450255

[15] O. Zariski and P. Samuel, *Commutative algebra, Volume I*, The University Series in Higher Mathematics, D. Van Nostrand Company, Inc., Princeton, New Jersey, 1958. With the cooperation of I. S. Cohen. MR0090581

THE INSTITUTE OF MATHEMATICAL SCIENCES, C.I.T CAMPUS, TARAMANI, CHENNAI - 600113, INDIA

Email address: anujjakhar@iisermohali.ac.in

INDIAN INSTITUTE OF SCIENCE EDUCATION AND RESEARCH MOHALI, SECTOR 81, KNOWLEDGE CITY, SAS NAGAR, PUNJAB - 140306, INDIA

Current address: Department of Mathematics, Panjab University, Chandigarh - 160014, India

Email address: skhanduja@iisermohali.ac.in

INDIAN INSTITUTE OF TECHNOLOGY, BOMBAY, POWAI, MUMBAI, MAHARASHTRA-400076, INDIA

Email address: neerajsan@iisermohali.ac.in

Contemporary Mathematics
Volume **738**, 2019
https://doi.org/10.1090/conm/738/14880

On Examples of Baer and Rickart Module Hulls

Jae Keol Park and S. Tariq Rizvi

Dedicated to the memory of Professor S.M. Abul Kazim Rizvi (1918–1980)

ABSTRACT. Since the discovery of injective hulls of modules, it has been of much interest to study various types of module hulls related to injectivity or its generalizations. These hulls allow us to work with modules having better properties and have been useful in obtaining the structures of modules or of the underlying rings. On the other hand, it is generally difficult to locate or construct explicit examples of various module hulls (including even injective hulls). In this paper, our focus is to construct and present explicit examples of Baer and Rickart module hulls, namely the smallest overmodules of a module M in a fixed injective hull of M with the respective properties mentioned. The study of Baer (and Rickart) rings was initiated by Kaplansky from a characterization of an AW^*-algebra as well as from the endomorphism ring of a semisimple module. The notions of Baer and Rickart rings have recently been extended to module theoretic notions. We discuss why the definitions of these two module notions are designed the way they are. In the main section of this expository paper, we discuss Baer and Rickart module hulls and their properties. Results which allow us to help construct explicit examples of Baer and Rickart module hulls or to present counterexamples related to them are included to make this paper as self-contained as possible. A number of motivating examples for further study are also included.

1. Introduction

The notion of AW^*-algebras was defined by Kaplansky [**10**] for the study of abstract W^*-algebras (also called von Neumann algebras). As a useful characterization of an AW^*-algebra, Kaplansky showed that a C^*-algebra A is an AW^*-algebra if and only if for every $\emptyset \neq X \subseteq A$, $r_A(X) = eR$ for some projection $e \in A$ (i.e., $e^2 = e$ and $e^* = e$), where $r_A(X)$ is the right annihilator of X in A and $*$ is the involution of A. Kaplansky then introduced the notion of a Baer ring in [**11**]. A ring R is called a Baer ring if the right annihilator of every nonempty subset of R is eR with $e^2 = e \in R$. It can be routinely checked that a ring R is a Baer ring if and only if the left annihilator of any nonempty subset of R is Rf with $f^2 = f \in R$, hence the notion is left-right symmetric. Kaplansky and many authors obtained a number of interesting results on Baer rings.

2010 *Mathematics Subject Classification.* 16D25, 16D40, 16D80, 16E60, 16R20, 16U10.

Key words and phrases. Baer module, Baer hull, Rickart module, Rickart hull, quasi-retractable.

The second author is the corresponding author.

Maeda [**20**] defined a Rickart ring. He called a ring right (resp., left) Rickart if the right (resp., left) annihilator of any single element is generated by an idempotent. Hattori [**9**] introduced the notion of a right PP ring, namely a ring in which every principal right ideal is projective. It was later discovered that a right Rickart ring is precisely a right PP ring.

More recently, the notion of a Baer ring was extended to an analogous module theoretic notion via the endomorphism ring of a module by Rizvi and Roman [**26**].

A module M_R is called a *Baer module* if, for any $N_R \leq M_R$, there is $e^2 = e \in S := \text{End}(M_R)$ with $\ell_S(N) = Se$, where $\ell_S(N) = \{f \in S \mid f(N) = 0\}$. A module M is said to be a *Rickart module* if for every $\phi \in \text{End}(M)$, $\text{Ker}(\phi) \leq^{\oplus} M$.

In Section 2, we discuss the design of the definitions of these modules. We note that a ring R is a Baer ring if and only if R_R is a Baer module. Similarly, a ring R is right Rickart if and only if R_R is a Rickart module. There are close connections between an extending ring and a Baer ring (see [**6**]). In [**26**] it was shown that every $\mathcal{K}$-nonsingular extending module is Baer and that every Baer module under a dual condition becomes extending. A number of interesting properties of Baer and Rickart modules are shown in [**16**], [**17**], [**26**], [**27**], and [**28**].

Let M_R be a module. We fix an injective hull $E(M_R)$ of M_R. Let $\mathfrak{M}$ be a class of modules. We call, when it exists, a module H_R the $\mathfrak{M}$ *hull* of M_R if H_R is the smallest extension of M_R in $E(M_R)$ that belongs to $\mathfrak{M}$ (see also [**2**, Definition 8.4.1]).

For a given module M, the smallest Baer (resp., Rickart) overmodule of M in $E(M)$ is called the Baer (resp., Rickart) hull of M. One of the difficulties in dealing with the Baer and the Rickart hull of M_R is the interplay of the scalar multiplication of M with R on one side of M and with the endomorphism ring $S = \text{End}(M_R)$ acting on the other side of M. Such an overmodule of M not only has to satisfy the conditions for being Baer (resp., Rickart) but also being the smallest such overmodule of M to be a hull. The Baer (resp., Rickart) property of the hull thus necessitates a consideration of the endomorphism rings of all overmodules of M in $E(M)$ before we can locate the smallest Baer (resp., Rickart) overmodule of M.

The description and construction of explicit examples of various types of hulls (including injective hulls) is difficult in general. In this expository paper, our focus is on presenting various explicit examples and explicit constructions of Baer hulls and Rickart hulls of certain classes of modules. To make this paper as self-contained as possible, we include results (without proofs) from various sources which allows us to construct these hulls or to build counterexamples. These examples will illustrate results that have been obtained so far to present the latest state of available literature in this regard. We expect this to help motivate further research on these topics.

All rings are assumed to have identity and all modules are assumed to be unitary. For right R-modules M_R and N_R, we use $\text{Hom}(M_R, N_R)$ or $\text{Hom}_R(M, N)$ to denote the set of all R-module homomorphisms from M_R to N_R. Likewise, $\text{End}(M_R)$ or $\text{End}_R(M)$ denotes the endomorphism ring of an R-module M. For an R-module homomorphism $f \in \text{Hom}_R(M, N)$, $\text{Ker}(f)$ denotes the kernel of f.

We use $E(M_R)$ or $E(M)$ to denote an injective hull of a module M_R. For a module M, $K \leq M$, $L \trianglelefteq M$, $N \leq^{\text{ess}} M$, and $U \leq^{\oplus} M$ denote that K is a submodule of M, L is a fully invariant submodule of M, N is an essential submodule of M, and U is a direct summand of M, respectively.

Let M be a right R-module. Then we let $r_R(X) = \{a \in R \mid Xa = 0\}$ for $\emptyset \neq X \subseteq M$. We use $\text{Ann}_R(M)$ for $r_R(M)$. Similarly, when M is a left S-module, we denote $\ell_S(Y) = \{s \in S \mid sY = 0\}$ for $\emptyset \neq Y \subseteq M$.

Let R be a ring and $\emptyset \neq X \subseteq R$, we let $r_R(X) = \{a \in R \mid Xa = 0\}$ and $\ell_R(X) = \{a \in R \mid aX = 0\}$ which are called the right annihilator of X in R and the left annihilator of X in R, respectively.

For a module M and a nonempty set Λ, let $M^{(\Lambda)}$ denote the direct sum of $|\Lambda|$ copies of M. When Λ is finite with $|\Lambda| = n$ and $n > 1$, $M^{(n)}$ stands for $M^{(\Lambda)}$. We use $\text{CFM}_\Lambda(R)$ to denote the $\Lambda \times \Lambda$ column finite matrix ring over a ring R. Let $\text{Mat}_n(R)$ denote the $n \times n$ matrix ring over a ring R.

For a ring R, we let $Q(R)$ stand for the maximal right ring of quotients of R. The symbols $\mathbb{C}$, $\mathbb{R}$, $\mathbb{Q}$, $\mathbb{Z}$, and $\mathbb{Z}_n (n > 1)$ denote the field of complex numbers, the field of real numbers, the field of rational numbers, the ring of integers, and the ring of integers modulo n, respectively. Ideals of a ring without the adjective "left" or "right" mean two-sided ideals. For a ring R, the notation $I \trianglelefteq R$ denotes that I is an ideal of R.

Assume that M and N are right R-modules such that $M = M_1 \oplus M_2 \oplus M_3$ and $N = N_1 \oplus N_2$. Let $S = \text{Hom}_R(M, N) = \text{Hom}_R(M_1 \oplus M_2 \oplus M_3, N_1 \oplus N_2)$. Then

$$S = \begin{bmatrix} \text{Hom}_R(M_1, N_1) & \text{Hom}_R(M_2, N_1) & \text{Hom}_R(M_3, N_1) \\ \text{Hom}_R(M_1, N_2) & \text{Hom}_R(M_2, N_2) & \text{Hom}_R(M_3, N_2) \end{bmatrix}.$$

In this case, for

$$\begin{bmatrix} f_{11} & f_{12} & f_{13} \\ f_{21} & f_{22} & f_{13} \end{bmatrix} \in S \text{ and } \begin{bmatrix} m_1 \\ m_2 \\ m_3 \end{bmatrix} \in M_1 \oplus M_2 \oplus M_3,$$

where $f_{ij} \in \text{Hom}_R(M_j, N_i)$ and $m_j \in M_j$ with $1 \leq i \leq 2$ and $1 \leq j \leq 3$,

$$\begin{bmatrix} f_{11} & f_{12} & f_{13} \\ f_{21} & f_{22} & f_{23} \end{bmatrix} \begin{bmatrix} m_1 \\ m_2 \\ m_3 \end{bmatrix} = \begin{bmatrix} f_{11}(m_1) + f_{12}(m_2) + f_{13}(m_3) \\ f_{21}(m_1) + f_{22}(m_2) + f_{23}(m_3) \end{bmatrix} \in N_1 \oplus N_2.$$

In general, $\text{Hom}_R(M, N)$ is described similarly when $M = M_1 \oplus \cdots \oplus M_n$ and $N = N_1 \oplus \cdots \oplus N_k$.

2. Baer and Rickart Modules

In this section, we begin with a historical background of the notion of Baer rings derived from AW^*-algebras by Kaplansky [**10**]. Some useful properties of Baer rings and Rickart rings are included. Based on the notion and important properties of Baer rings, designing the notion of Baer modules in a module theoretic setting is discussed. In particular, we highlight designing the notion of a Baer module definition from the notion of Baer rings. The design of the notion of a Rickart module definition from the notion right Rickart ring is shown.

Basic properties and some useful results on Baer and Rickart modules to be used in Section 3 are provided. Examples of Baer and Rickart rings and modules are also discussed to illustrate the notions.

Assume that R is a ring. Recall that a map $*: R \to R$ is called an involution if (i) $(x+y)^* = x^* + y^*$; (ii) $(xy)^* = y^*x^*$; and (iii) $(x^*)^* = x$ for all $x, y \in R$. A ring R with an involution $*$ is called a *$*$-ring*. An element $e \in R$ is called a *projection* if $e^2 = e$ and $e^* = e$ (i.e., e is a self-adjoint idempotent).

Recall that a Banach algebra is a complex normed algebra A which is complete (as a topological space) such that $||ab|| \leq ||a||\,||b||$ for all $a, b \in A$. A *Banach $*$-algebra* is a complex Banach algebra A with an involution $*$ satisfying that $(\lambda a)^* = \overline{\lambda}\, a^*$ for $\lambda \in \mathbb{C}$ and $a \in A$, where $\overline{\lambda}$ is the conjugate of $\lambda \in \mathbb{C}$.

A *C^*-algebra* A is a Banach $*$-algebra with the additional norm condition that $||a^*a|| = ||a||^2$ for all $a \in A$. A C^*-algebra is called unital if it has identity.

By Kaplansky [**10**], the notion of AW^*-algebras was introduced for the study of abstract W^*-algebras (also called von Neumann algebras).

DEFINITION 2.1 ([**10**, Kaplansky]). An AW^*-algebra A is a C^*-algebra which satisfies the following.

(i) In the partially ordered set of projections of A, any set of orthogonal projections has a least upper bound.

(ii) Any maximal commutative self-adjoint subalgebra B (i.e., $B^* = B$) is generated by its projections (that is, B is equal to the smallest closed subalgebra containing its projections).

It is shown in [**10**] that an AW^*-algebra is a unital C^*-algebra. There are several characterizations of an AW^*-algebra. One of the most useful characterizations is as follows (see [**10**] and [**3**, p.243]).

THEOREM 2.2 ([**10**, Kaplansky]). *A C^*-algebra A is an AW^*-algebra if and only if the left annihilator of any nonempty subset of A is Ae for some projection e of A if and only if the right annihilator of any nonempty subset of A is fA for some projection f of A.*

The next theorem, essentially due to Baer [**1**], provides suggestions for defining the notion of a Baer ring (Kaplansky [**11**]).

THEOREM 2.3. *Assume that M_R is a semisimple R-module and $S = End(M_R)$. Then the left annihilator of any nonempty subset of S is Se for some $e^2 = e \in S$. Further, the right annihilator of any nonempty subset of S is also fS for some $f^2 = f \in S$.*

THEOREM 2.4 ([**11**, Theorem 3, p.2]). *In a ring R, not necessarily with identity, any two of the following statements imply the third.*

(i) For each $\emptyset \neq X \subseteq R$, there exists $e^2 = e \in R$ such that $\ell_R(X) = Re$.

(ii) For each $\emptyset \neq Y \subseteq R$, there exists $f^2 = f \in R$ such that $r_R(Y) = fR$.

(iii) R has an identity.

From Theorems 2.3 and 2.4, Kaplansky [**11**] defined the notion of Baer rings as follows in the ring theoretical setting.

DEFINITION 2.5. A ring R is called *Baer* if R satisfies any two (hence, all three) of the statements of Theorem 2.4.

From Theorem 2.4, any Baer ring always has an identity, and the conditions for a ring to be Baer is left-right symmetric. Let R be a $*$-ring. Then R is called a Baer $*$-ring if and only if $r_R(X) = eR$ with $e \in R$ a projection for every $\emptyset \neq X \subseteq R$.

Thus by Theorem 2.2, a C^*-algebra A is an AW^*-algebra if and only if A is a Baer $*$-ring.

EXAMPLE 2.6. (i) Every AW^*-algebra is a Baer ring.

(ii) The endomorphism ring of a semisimple module is a Baer ring (see Theorem 2.3).

(iii) The ring of all sequences of complex numbers which are eventually real numbers is a Baer ring.

(iv) Any right self-injective von Neumann regular ring is a Baer ring.

(v) Let R be a domain and n be an integer such that $n > 1$. Then $T_n(R)$ is a Baer ring if and only if R is a division ring ([**11**, Exercises 2, p.16]), where $T_n(R)$ is the $n \times n$ upper triangular matrix ring over R.

In [**20**], Maeda defined Rickart rings as in the following definition. Hattori [**9**] introduced the notion of a right PP ring (i.e., a ring with the property that every principal right ideal is projective). It was later shown that right PP rings are precisely right Rickart rings (Example 2.8(v)).

DEFINITION 2.7. A ring R is called *right Rickart* if the right annihilator of any element is generated by an idempotent. A left Rickart ring is defined similarly.

EXAMPLE 2.8. (i) Every von Neumann regular ring is a right (and left) Rickart ring.

(ii) Every Baer ring is a right (and left) Rickart ring.

(iii) The center of a right Rickart ring is a right (and left) Rickart ring.

(iv) Let $A_n = \mathbb{Z}$ for $n = 1, 2, \ldots$, and put

$$R = \left\{ (a_n)_{n=1}^{\infty} \in \prod_{n=1}^{\infty} A_n \mid a_n \text{ is constant eventually} \right\},$$

a subring of $\prod_{n=1}^{\infty} A_n$. Then R is a Rickart ring. Indeed, say $\alpha = (a_n)_{n=1}^{\infty} \in R$. Let $e = (e_n)_{n=1}^{\infty}$, where $e_n = 1$ if $a_n = 0$ and $e_n = 0$ if $a_n \neq 0$. Since a_n is constant eventually, so is e_n. Also $e^2 = e \in R$ and $r_R(\alpha) = eR$. But R is neither Baer nor von Neumann regular.

(v) Recall that a ring R is said to be right PP if every principal right ideal of R is projective as a right R-module. A ring R is right Rickart if and only if R is right PP. Hence every right hereditary ring is right Ricakrt.

(vi) A ring R is called right (resp., left) hereditary if every right (resp., left) ideal of R is projective as a right (resp., left) R-module. Any right (resp., left) hereditary ring is a right (resp., left) Rickart ring.

(vii) By Chase [**4**], there is a ring R such that R is left hereditary (hence left Rickart), but R is not right Rickart (see also [**2**, Example 3.1.20] and [**5**, Example 6.21]). Take $F_n = \mathbb{Z}_2$ for $n = 1, 2, \ldots$. Let

$$A = \left\{ (a_n)_{n=1}^{\infty} \in \prod_{n=1}^{\infty} F_n \mid a_n \text{ is constant eventually} \right\},$$

a subring of $\prod_{n=1}^{\infty} F_n$, and let $I = \{(a_n)_{n=1}^{\infty} \in \prod_{n=1}^{\infty} F_n \mid a_n = 0 \text{ eventually}\}$, which is an ideal of A. Put $S = A/I$ and

$$R = \begin{bmatrix} S & S \\ 0 & A \end{bmatrix}.$$

Then R is left hereditary. Hence R is a left Rickart ring. Say $\alpha = \begin{bmatrix} 0 & 1_S \\ 0 & 0 \end{bmatrix} \in R$, where 1_S is the identity of S. Then $r_R(\alpha)$ is not generated by an idempotent of R. Therefore R is not right Rickart.

A ring R is said to be right semihereditary if every finitely generated right ideal of R is projective as a right R-module. A commutative semihereditary domain is called a Prüfer domain.

A module M is called extending if every submodule of M is essential in a direct summand of M. A ring R is called right (resp., left) extending if the module R_R (resp., ${}_RR$) is extending.

THEOREM 2.9 (see [**2**, Theorem 6.1.4]). *Let R be a commutative domain. Then the following are equivalent.*

(i) R is a Prüfer domain.

(ii) $Mat_n(R)$ is a Baer (Rickart) ring for every positive integer n.

(iii) $Mat_k(R)$ is a Baer (Rickart) ring for some integer $k > 1$.

(iv) $Mat_2(R)$ is a Baer (Rickart) ring.

Furthermore, in (ii), (iii), and (iv), "Baer (Rickart) ring" can be can be replaced by "(right) extending ring".

The following theorem is due to Lenzing [**19**, Satz 3]. Recall that $\mathrm{CFM}_\Lambda(R)$ denotes the $\Lambda \times \Lambda$ column finite matrix ring over a ring R.

THEOREM 2.10. *Let R be a ring. Then R is right hereditary if and only if $CFM_\Lambda(R)$ is right Rickart for any nonempty set Λ.*

We now discuss how to design the notion of a Baer module in a general module theoretic setting by using the endomorphism ring of a module. Then useful and interesting properties of Baer modules are discussed. Moreover, connections between Baer modules and extending modules are shown.

Let M be a right R-module and $S = \mathrm{End}(M_R)$. Because the Baer ring property is left-right symmetric (see Theorem 2.4 and Definition 2.5), we have the following two candidates for a suitable definition of the notion of Baer modules from the definition of Baer rings:

Candidate **(A)**: For each $\emptyset \neq X \subseteq M$, $\ell_S(X) = Se$ for some $e^2 = e \in S$, where $\ell_S(X) = \{f \in S \mid fX = 0\}$.

Candidate **(B)**: For each $\emptyset \neq X \subseteq M$, $r_R(X) = fR$ for some $f^2 = f \in R$, where $r_R(X) = \{a \in R \mid Ma = 0\}$.

For a suitable definition of the notion of Baer modules, first we expect the following:

(1) The endomorphism ring S is a Baer ring so that the type theory of S can be applied to a decomposition of M.

In view of **(1)** and Theorem 2.3, we also have the following expectation:

(2) The notion of Baer modules needs to be designed so that every semisimple module should be a Baer module.

The following provides a key motivation for a suitable definition of the notion of a Baer module.

THEOREM 2.20. *Assume that M_R is a semisimple R-module and $S = End(M_R)$. Then the conditions for Candidate* **(A)** *are satisfied.*

PROOF. Take $\emptyset \neq X \subseteq M$. Put $N = XR$. Then $N_R \leq^{\oplus} M_R$ as M is semisimple. So $M = N \oplus K$ for some $K \leq M$. Hence

$$S = \text{End}_R(N \oplus K) = \begin{bmatrix} \text{End}_R(N) & \text{Hom}_R(K,N) \\ \text{Hom}_R(N,K) & \text{End}_R(K) \end{bmatrix}.$$

Let $s \in \ell_S(N)$. Then

$$s = \begin{bmatrix} \alpha & \beta \\ \gamma & \delta \end{bmatrix} \in S$$

with $\alpha \in \text{End}_R(N)$, $\beta \in \text{Hom}_R(K,N)$, $\gamma \in \text{Hom}_R(N,K)$ and $\delta \in \text{End}_R(K)$ such that

$$0 = \begin{bmatrix} \alpha & \beta \\ \gamma & \delta \end{bmatrix} \begin{bmatrix} N \\ 0 \end{bmatrix} = \begin{bmatrix} \alpha(N) \\ \gamma(N) \end{bmatrix}.$$

So $\alpha = 0$ and $\gamma = 0$, and thus

$$\ell_S(N) = \begin{bmatrix} 0 & \text{Hom}_R(K,N) \\ 0 & \text{End}_R(N) \end{bmatrix}.$$

Let $1_{\text{End}_R(N)}$ be the identity of $\text{End}_R(N)$, and let

$$e = \begin{bmatrix} 0 & 0 \\ 0 & 1_{\text{End}_R(N)} \end{bmatrix} \in S.$$

Then $e^2 = e$ and $\ell_S(N) = Se$. Note that $\ell_S(X) = \ell_S(XR) = \ell_S(N)$ because $N = XR$. Therefore $\ell_S(X) = Se$ with $e^2 = e \in S$. □

By Theorem 2.20, any semisimple module satisfies conditions listed for Candidate **(A)**. Furthermore, the following also should be counted for designing the notion of Baer modules.

Let M_R be a right R-module and $S = \text{End}_R(M)$. Put $M^* = \text{Hom}(M_R, R_R)$. Then M is an (S,R)-bimodule and M^* is an (R,S)-bimodule. Consider an (R,R)-bimodule homomorphism $(-,-)$ and an (S,S)-bimodule homomorphism $[-,-]$:

$$(-,-) : {}_RM^* \otimes_S M_R \to R \text{ with } (f,m) = f(m)$$

for $f \in M^*$ and $m \in M$, and

$$[-,-] : {}_SM \otimes_R M^*_S \to S \text{ with } [m,f]n = mf(n)$$

for $m, n \in M$ and $f \in M^*$.

Then $m(f,n) = [m,f]n$ for $m, n \in M$ and $f \in M^*$. Moreover, $(-,-)$ is an (R,R)-bimodule homomorphism, while $[-,-]$ is an (S,S)-bimodule homomorphism. Hence there exists a Morita context $(R, {}_SM_R, {}_RM^*_S, S)$ derived from the module M_R with the pairrings $(-,-)$ and $[-,-]$. See [**14**, Section 18C] for more details on Morita context.

(3) The important role that the endomorphism ring S of a right R-module M plays should be considered in view of the Morita context induced from M because S is one of key ingredients of the Morita context derived from M_R. Thereby, for a suitable notion of a Baer module, it is necessary to keep in mind that ${}_SM_R$ is an (S,R)-bimodule so that S can provide useful information for the study of M_R through interaction between M_R and S.

(4) The notion of Baer rings is left-right symmetric. Thus some module theoretic analogue of the left-right symmetry property of the notion of Baer modules should happen.

REMARK 2.21. In [**18**, Definition 2.1], Lee and Zhou also called a module M_R *Baer* if M satisfies the condition of Candidate **(B)**. In other words, M is Baer if, for any nonempty subset X of M, $r_R(X) = eR$ with $e^2 = e \in R$.

EXAMPLE 2.22. (i) Let $M_R = \mathbb{Z}_p$ (p a prime integer) and $R = \mathbb{Z}$. Put $Y = M$. Then we see that $r_R(Y) = p\mathbb{Z}$, thus there is no $e^2 = e \in R$ such that $r_R(Y) = eR$. We note that M is a simple R-module. But M does not satisfy the conditions of Candidate **(B)**.

(ii) In general, let R be a commutative domain. If there exists a simple R-module satisfying the conditions of Candidate **(B)**, then R should be field. In fact, say $M := R/I$ is a simple R-module with the conditions of Candidate **(B)**. If R is not a field, then $I \neq 0$ since the ideal I is maximal. Note that $r_R(M) = I$, and thus $I = eR$ for some $e^2 = e \in R$. Hence $I = 0$ or $I = R$. So $I = R$, which is a contradiction.

Consequently, if R is a commutative domain which is not a field, then there is no simple R-module satisfying the conditions of Candidate **(B)**.

EXAMPLE 2.23. Let $M_R = \mathbb{Z}[x] \oplus \mathbb{Z}[x]$ and $R = \mathbb{Z}[x]$. Then for any nonempty subset Y of M, $r_R(Y) = fR$ for some $f^2 = f \in R$ (in fact, $f = 0$ or $f = 1$). So M satisfies the conditions for Candidate **(B)**. However, $\text{End}(M_R) = \text{Mat}_2(\mathbb{Z}[x])$ is not a Baer ring by Theorem 2.9 because $\mathbb{Z}[x]$ is not Prüfer.

Thus in view of **(1)**, **(2)**, **(3)**, **(4)**, Theorem 2.20, Example 2.22, and Example 2.23, we adopt conditions of Candidate **(A)** to define a Baer module as follows:

DEFINITION 2.24 (see [**26**, Definition 2.2]). Let M be a right R-module with $S = \text{End}(M_R)$. Then M is called a *Baer module* if conditions listed for Candidate **(A)** are satisfied. That is, for any $\emptyset \neq X \subseteq M$, $\ell_S(X) = Se$ such that $e^2 = e \in S$.

Let M be a right R-module and $\emptyset \neq X \subseteq M$. We put $N = XR \leq M$. Then $\ell_S(X) = \ell_S(XR) = \ell_S(N)$. Hence Definition 2.24 is equivalent to the following definition.

DEFINITION 2.25 ([**26**, Definition 2.2]). Let M be a right R-module with $S = \text{End}(M_R)$. Then M is called a *Baer module* if for any $N \leq M$, $\ell_S(N) = Se$ for some $e^2 = e \in S$.

Assume that M_R is a module with $S = \text{End}(M_R)$. For any $\emptyset \neq T \subseteq S$, we let $r_M(T) = \{m \in M \mid Tm = 0\}$.

The Baer property of rings is left-right symmetric. In view of **(3)** and **(4)**, the following result shows a module theoretic analogue of this fact.

THEOREM 2.26 (see [**26**, Lemma 1.9 and Remark 2.3]). *The following are equivalent for a right R-module M.*

(i) M is a Baer module.

(ii) For any left ideal I of S, $r_M(I) = fM$ with $f^2 = f \in S$.

(iii) For any $\emptyset \neq Y \subseteq S, r_M(Y) = gM$ with $g^2 = g \in S$.

Let R be a ring. Then R is a Baer ring if and only if R_R is a Baer module from Theorem 2.26. Further, Theorem 2.20 and Definition 2.25 yield that all semisimple modules are Baer modules. Several other examples of Baer modules will follow later.

THEOREM 2.27 ([**26**, Theorem 4.1]). *Let M_R be a Baer module. Then End(M_R) is a Baer ring.*

The following example exhibits that the converse of Theorem 2.27 is not true.

EXAMPLE 2.28. Let $M_R = \mathbb{Z}_{p^\infty}$ (p is a prime integer), the Prüfer p-group, and $R = \mathbb{Z}$. Then $\text{End}(M_R) = \widehat{\mathbb{Z}}_p$, the domain of p-adic integers. Thus $\text{End}(M_R)$ is a Baer ring, but M_R is not a Baer module.

From Example 2.28, for characterizing a Baer module in terms of its endomorphism ring, an additional assumption is necessary. For this, the following notion is introduced which is defined by Rizvi and Roman [**28**].

DEFINITION 2.29 ([**28**, Definition 2.3]). Let M be a right R-module and $S = \text{End}(M_R)$. Then M is called quasi-retractable if $\text{Hom}_R(M, r_M(I)) \neq 0$ for every left ideal I of S with $r_M(I) \neq 0$.

We remark that a module M_R is quasi-retractable if and only if $r_S(I) \neq 0$ for every left ideal I of S with $r_M(I) \neq 0$.

Recall that a module M is said to be *retractable* if $\text{Hom}_R(M, N) \neq 0$, for any $0 \neq N \leq M$. Examples of retractable modules include free modules, generators, and semisimple modules. Obviously, retractable modules are quasi-retractable.

In the following, we provide some additional classes of quasi-retractable modules.

THEOREM 2.30 ([**24**, Theorem 2.3]). *(i) If R is a ring and A is an intermediate ring between R and $Q(R)$, then $A_R^{(n)}$ is quasi-retractable for any positive integer n.*

(ii) If R is a prime PI-ring (e.g., R is a commutative domain) and A is an intermediate ring between R and $Q(R)$, then $A_R^{(\Lambda)}$ is quasi-retractable for any nonempty set Λ. Furthermore, End($A_R^{(\Lambda)}$) = End($A_A^{(\Lambda)}$) = $CFM_\Lambda(A)$.

(iii) If R is a right nonsingular ring and M is an intermediate (R, R)-bimodule between R and $Q(R)$, then $M_R^{(n)}$ is quasi-retractable for any positive integer n.

REMARK 2.31 ([**24**, Remark 2.4]). Let R be a commutative domain with the field of fractions F. For an intermediate domain A between R and F, $A_R^{(\Lambda)}$ is quasi-retractable for any nonempty set Λ by Theorem 2.30(ii). But $A_R^{(\Lambda)}$ may not be retractable. For example, $\mathbb{Q}_\mathbb{Z}$ is quasi-retractable. But $\mathbb{Q}_\mathbb{Z}$ is not retractable because $\text{Hom}(\mathbb{Q}_\mathbb{Z}, \mathbb{Z}_\mathbb{Z}) = 0$.

The following result, due to Rizvi and Roman [**28**], is a full characterization of a Baer module via its endomorphism ring and the quasi-retractability.

THEOREM 2.32 ([**28**, Theorem 2.5]). *A module M is a Baer module if and only if S = End(M) is a Baer ring and M is quasi-retractable.*

The next theorem is useful for the study of eR_R when R is a Baer ring and $e^2 = e \in R$.

THEOREM 2.33 ([**26**, Theorem 2.17]). *Every direct summand of a Baer module is a Baer module.*

Hence given a Baer ring R, one automatically obtains that the right R-module eR_R (where $0 \neq e^2 = e \in R$) inherits the module theoretic Baer property. Theorem 2.27 and Theorem 2.33 yield the following corollary immediately.

COROLLARY 2.34. *Let R be a Baer ring and $0 \neq e^2 = e \in R$. Then we have the following.*
(i) eR_R is a Baer module.
(ii) End(eR_R) $= eRe$ is a Baer ring.

DEFINITION 2.35 ([**26**, Definition 2.5]). Let M_R be a module and $S = \text{End}(M_R)$. Then M is called $\mathcal{K}$-nonsingular if, for $\phi \in S$, $r_M(\phi) = \text{Ker}(\phi) \leq^{\text{ess}} M$ implies $\phi = 0$.

The following result is [**26**, Proposition 2.10].

PROPOSITION 2.36. *Every nonsingular module is $\mathcal{K}$-nonsingular.*

The following example shows that the converse of Proposition 2.36 does not hold true. But when $M = R$, the $\mathcal{K}$-nonsingularity and the nonsingularity of M coincide. For more details on $\mathcal{K}$-nonsingular modules, see [**27**].

EXAMPLE 2.37. The $\mathbb{Z}$-module $\mathbb{Z}_p$ (p is a prime integer) is $\mathcal{K}$-nonsingular. However, $\mathbb{Z}_p$ is not nonsingular.

DEFINITION 2.38 ([**26**, Definition 2.7]). A right R-module M with $S = \text{End}(M_R)$ is called $\mathcal{K}$-cononsingular if, for $N \leq M$, $\ell_S(N) = 0$ implies $N \leq^{\text{ess}} M$.

By Chatters and Khuri [**6**], a right extending ring which is right nonsingular is precisely a Baer and right cononsingular ring. For a module theoretic analogue of this result, we have the following due to Rizvi and Roman [**26**], which shows strong connections between extending modules and Baer modules.

THEOREM 2.39 ([**26**, Theorem 2.12]). *Let M_R be a module. Then the following are equivalent.*
(i) M_R is extending and $\mathcal{K}$-nonsingular.
(ii) M_R is Baer and $\mathcal{K}$-cononsingular.

In general, a direct sum of Baer modules is not always Baer. On the other hand, the next result shows an instance where this holds true. The following result [**2**, Theorem 4.2.18] is useful for the study of module hulls.

THEOREM 2.40. *Let V be a nonsingular injective module and M be a nonsingular extending module. Then $V \oplus M$ is a Baer module.*

Next, we discuss a suitable design of the notion of a Rickart module. Note that any Baer ring is a right Rickart ring. Thus for any suitable design of the notion of a Rickart module, we expect that if M_R is a Baer module, then M_R should be a Rickart module. Thereby, any semisimple module should also be a Rickart module because any semisimple module is a Baer module (see Theorem 2.20).

Furthermore, by definition of a Baer module, a ring R is a Baer ring if and only if R_R is a Baer module. Thus for a ring R, it is natural that the notion should be designed so that R is a right Rickart ring if and only if R_R is a Rickart module.

We let M_R be a right R-module and $S = \text{End}(M_R)$. Since the notion of a Rickart module is designed so that any Baer module is a Rickart module, the following may be conditions for a Rickart module candidate suggested from Definition 2.24.

Candidate **(C)**: For each $m \in M$, $\ell_S(m) = Se$ for some $e^2 = e \in S$.

Suppose conditions listed for Candidate **(C)** are adopted for the definition of a Rickart right R-module, that is, a module M_R is Rickart if and only if for each $m \in M$, $\ell_S(m) = Se$ for some $e^2 = e \in S$, where $S = \text{End}(M_R)$.

Assume that R is a right Rickart ring. If R_R is a Ricakrt module, then since $S = \text{End}(R_R) = R$, conditions listed for Candidate **(C)** imply that R is a left Rickart ring. Thus every right Rickart ring is a left Rickart ring. By modification of Example 2.8(vii), there exists a right Rickart ring which is not a left Rickart ring. So we get a contradiction. Hence R_R is not a Rickart module. So conditions listed for Candidate **(C)** *disqualify it* to be a suitable definition of a Rickart right R-module.

Next, consider the following Candidate **(D)** for the notion of a right Right Rickart module:

Candidate **(D)**: For any $m \in M$, $r_R(m) = eR$ for some $e^2 = e \in R$.

Assume that conditions listed for Candidate **(D)** are adopted for the notion of right Rickart modules, that is, a right R-module M_R is Rickart if and only if for any $m \in M$, $r_R(m) = eR$ for some $e^2 = e \in R$. Let $M = \mathbb{Z}_p$ (p is a prime integer). As a $\mathbb{Z}$-module, M is a Baer module since it is a simple $\mathbb{Z}$-module. So we expect that M is Rickart as a $\mathbb{Z}$-module. But it is not. Indeed, let $m = \overline{1}$ be the image of 1 in $\mathbb{Z}_p$. Then $r_{\mathbb{Z}}(m) = p\mathbb{Z}$. So there is no $e^2 = e \in \mathbb{Z}$ such that $r_{\mathbb{Z}}(m) = e\mathbb{Z}$. Therefore M is not Rickart as a $\mathbb{Z}$-module. Thus conditions listed for Candidate **(D)** are also *disqualified* to be adopted as the definition of Rickart module.

Instead of the non-suitable Candidates **(C)** and **(D)**, we now consider the following conditions suggested by condition (iii) of Theorem 2.26.

Candidate **(E)**: For any $\phi \in S$, $r_M(\phi) = eM$ for some $e^2 = e \in S$.

Before adopting Candidate **(E)** as the definition for the notion of a Rickart module, we observe that the following holds true.

PROPOSITION 2.41. *(i) Any Baer module M_R satisfies the conditions of Candidate* **(E)**.

(ii) A ring R is right Rickart if and only if R_R satisfies Candidate **(E)**.

From Proposition 2.41, adopting Candidate **(E)** for the definition of a Rickart module makes sense as the next definition shows.

DEFINITION 2.42 ([**16**, Definition 2.2]). A module M_R with $S = \text{End}(M_R)$ is called a *Rickart module* if, for any $\phi \in S$, $r_M(\phi) = eM$ for some $e^2 = e \in S$.

Note in Definition 2.42 that $r_M(\phi) = \text{Ker}(\phi)$. For a ring R, by Definition 2.42, R_R is a Rickart module if and only if R is a right Rickart ring. From Theorem 2.26 and Definition 2.42, any Baer module is a Rickart module.

The following results show that any direct summand and the endomorphism ring of a Rickart module inherit the Rickart property naturally and without any additional conditions.

THEOREM 2.43. *(i) ([**16**, Theorem 2.7]). Every direct summand of a Rickart module is a Rickart module.*

(ii) ([**16**, Proposition 3.2]). *The endomorphism ring of a Rickart right R-module is a right Rickart ring.*

Theorem 2.43 yields the following corollary immediately.

COROLLARY 2.44. *Let R be a right Rickart ring and $e^2 = e \in R$. Then we have the following.*
(i) eR_R is a Rickart module.
(ii) eRe is a right Rickart ring.

THEOREM 2.45 ([**16**, Proposition 3.5]). *Assume that M_R is a quasi-retractable R-module. Then M_R is a Rickart module if and only if End(M_R) is a right Rickart ring.*

Every free module is retractable and so it is quasi-retractable. Using this fact, the following corollary is obtained.

COROLLARY 2.46. *Let R be a ring and Λ be a nonempty set. Then the following are equivalent.*
(i) $R_R^{(\Lambda)}$ is a Rickart module.
(ii) $CFM_\Lambda(R)$ is a right Rickart ring.

3. Baer Module Hulls, Rickart Module Hulls, and Examples

In this section, we provide results on Baer and Rickart module hulls. The included results help us construct various interesting examples and counterexamples related to Baer and Rickart hulls of modules. Several motivating examples are also included for further research.

DEFINITION 3.1 ([**2**, Definition 8.4.1]). Let M_R be a module. We fix an injective hull $E(M_R)$ of M_R. Let $\mathfrak{M}$ be a class of modules. We call, when it exists, a module H_R the $\mathfrak{M}$ *hull* of M_R if H_R is the smallest extension of M_R in $E(M_R)$ that belongs to $\mathfrak{M}$.

Assume that M is a module. By Definition 3.1, the *Baer* (resp., *Rickart,* etc.) *hull* of M is the smallest Baer (resp., Rickart, etc.) essential extension of M in $E(M)$. We use $\mathbf{B}(-)$ and $\mathbf{Ric}(-)$ to denote the Baer module hull and the Rickart module hull of a module, respectively when they exist. For our convenience, we will drop "module" in these "module hulls" of modules and call these just Baer hull or Rickart hull.

For a module M over a commutative domain R, we use $t(M)$ to denote the torsion submodule of M, that is,

$$t(M) = \{m \in M \mid ma = 0 \text{ for some nonzero } a \in R\}.$$

When $t(M) = 0$, M is called *torsion-free.* When $M = t(M)$, we say that M is a *torsion module.*

PROPOSITION 3.2 ([**15**, Proposition 2.7]). *Let N be a finitely generated $\mathbb{Z}$-module. Then the following are equivalent.*
(i) N is a Baer module.
(ii) N is a Rickart module.
(iii) N is semisimple or torsion-free.

See [**30**, 6.2.10, p.98] for the proof of the following theorem.

THEOREM 3.3 ([**13**, Kulikov]). *If G is an abelian group which is not torsion-free (as a $\mathbb{Z}$-module), then G has a direct summand which is either cyclic of prime power order or $\mathbb{Z}_{p^\infty}$ for some prime integer p, where $\mathbb{Z}_{p^\infty}$ is the Prüfer p-group.*

From Proposition 3.2. and Theorem 3.3, we have the following two motivating examples.

EXAMPLE 3.4. Let $N = \mathbb{Z}_p \oplus \mathbb{Z}$ (p a prime integer). Then as a $\mathbb{Z}$-module, N is not Baer (Rickart) by Proposition 3.2. Now $\mathbb{Z}_p$ and $\mathbb{Q}$ are Baer (Rickart) as $\mathbb{Z}$-modules, $\mathrm{Hom}_{\mathbb{Z}}(\mathbb{Z}_p, \mathbb{Q}) = 0$ and $\mathrm{Hom}_{\mathbb{Z}}(\mathbb{Q}, \mathbb{Z}_p) = 0$. Thus

$$\mathrm{End}_{\mathbb{Z}}(\mathbb{Z}_p \oplus \mathbb{Q}) = \mathrm{End}_{\mathbb{Z}}(\mathbb{Z}_p) \oplus \mathrm{End}_{\mathbb{Z}}(\mathbb{Q}) = \mathbb{Z}_p \oplus \mathbb{Q},$$

which is a Baer (Rickart) ring. Therefore it can be routinely checked that $\mathbb{Z}_p \oplus \mathbb{Q}$ is a Baer (Rickart) $\mathbb{Z}$-module. Thus $\mathbb{Z}_p \oplus \mathbb{Q}$ is a Baer (Rickart) essential extension of N.

Next, let V be an intermediate Baer (resp., Rickart) $\mathbb{Z}$-module between N and $E(N) = \mathbb{Z}_{p^\infty} \oplus \mathbb{Q}$, where $\mathbb{Z}_{p^\infty}$ is the Prüfer p-group. Then $\mathrm{udim}(V) = 2$ because $\mathrm{udim}(\mathbb{Z}_p \oplus \mathbb{Z}) = 2$, where $\mathrm{udim}(-)$ denotes the uniform dimension.

By Theorem 3.3, there exists a prime integer q such that $\mathbb{Z}_{q^k}$ is a direct summand of V for some positive integer k, or there is a prime integer q_1 such that $\mathbb{Z}_{q_1^\infty}$ is a direct summand of V. Hence either $V = \mathbb{Z}_{q^k} \oplus W$ for some $W \leq V$ or $V = \mathbb{Z}_{q_1^\infty} \oplus W_1$ for some $W_1 \leq V$.

If $V = \mathbb{Z}_{q_1^\infty} \oplus W_1$, then $\mathbb{Z}_{q_1^\infty}$ is Baer as a $\mathbb{Z}$-module by Theorem 2.33. But this is impossible. Hence $V = \mathbb{Z}_{q^k} \oplus W$. Then W is torsion-free by Theorem 3.3 because $\mathrm{udim}(V) = 2$. Since $N \leq V$, $\mathbb{Z}_p \leq \mathbb{Z}_{q^k}$. Hence $q = p$, and so $V = \mathbb{Z}_{p^k} \oplus W$.

Because V is a Baer module, $\mathbb{Z}_{p^k}$ is a Baer module from Theorem 2.33. Thus $k = 1$ by Proposition 3.2. Therefore $V = \mathbb{Z}_p \oplus W$ with W torsion-free. We note that $\mathbb{Z}_p \oplus \mathbb{Z} \leq \mathbb{Z}_p \oplus W \leq \mathbb{Z}_{p^\infty} \oplus \mathbb{Q}$. Consequently,

$$V = \mathbb{Z}_p \oplus W \quad \text{with } \mathbb{Z} \leq W \leq \mathbb{Q}.$$

Since $\mathbb{Z}_p \oplus \mathbb{Q}$ is a Baer (resp., Rickart) $\mathbb{Z}$-module, $\mathbb{Z}_p \oplus \mathbb{Q}$ is the largest Baer (resp., Rickart) $\mathbb{Z}$-module between $N = \mathbb{Z}_p \oplus \mathbb{Z}$ and $E(N)$.

EXAMPLE 3.5. Let $W = \mathbb{Z}_{p^k} \oplus \mathbb{Z}$ (p a prime integer) as a $\mathbb{Z}$-module, where k is an integer such that $k > 1$. Then as in the argument used in Example 3.4, there is no Baer(resp., Rickart) essential extension of W. Hence W has no smallest Baer (resp., Ricakrt) essential extension. In other words, W has no Baer (resp., Ricakrt) hull.

Let R be a commutative noetherian domain which is not a field and F be its field of fractions. Assume that $N = M_R \oplus (\oplus_{i \in \Lambda} K_i)_R$, where M is semisimple with only a finite number of homogeneous components, and $\{K_i \mid i \in \Lambda\}$ is a set of nonzero submodules of F_R. The following result describes all intermediate modules between N and $E(N)$ which happen to be Baer (Rickart) modules.

THEOREM 3.6 ([**21**, Theorem 2.6] and [**15**, Lemma 3.2]). *Let R be a commutative noetherian domain which is not a field. Assume that M is a semisimple R-module with only a finite number of homogeneous components, and $\{K_i \mid i \in \Lambda\}$ is a set of nonzero submodules of F_R, where F is the field of fractions of R. Let V_R be an essential extension of $M_R \oplus (\oplus_{i \in \Lambda} K_i)_R$. Then the following are equivalent.*

(i) V is a Baer (resp., Rickart) module.

(ii) (1) $V = M \oplus W$ for some Baer (resp., Rickart) essential extension W of $(\oplus_{i\in\Lambda} K_i)_R$.

(2) $Hom_R(W, M) = 0$.

Assume that R is a commutative domain with the field of fractions F. Let B be a nonzero ideal of R. We put $B^0 = R$. For each non-negative integer ℓ, and let

$$[R : B^\ell] = (B^\ell)^{-1} = \{q \in F \mid qB^\ell \subset R\}.$$

We let $T(B) = \bigcup_{\ell\geq 0} [R : B^\ell]$. Then

$$T(B) = \sum_{\ell\geq 0}[R : B^\ell] = \sum_{\ell\geq 0}(B^\ell)^{-1}$$

since $R = [R : B^0] \subseteq [R : B] \subseteq [R : B^2] \subseteq \dots$. We remark that $T(B)$ is an intermediate domain between R and F. The domain $T(B)$ is called the *Nagata transform* (or the *ideal transform*) of B (see [**7**, p.490] and [**8**, p.325]).

Let R be a commutative domain with the field of fractions F. For an ideal B of R, recall that $B^{-1} = \{q \in F \mid qB \subseteq R\}$. It is well-known that for a nonzero ideal I of a commutative domain R, I_R is projective if and only if $II^{-1} = R$. In this case, I_R is finitely generated.

Recall that a commutative domain R is called Dedekind if every ideal of R is projective as an R-module. Therefore a Dedekind domain is precisely a commutative hereditary domain. Hence for each nonzero ideal I of a Dedekind domain R, it follows that $II^{-1} = R$ because I_R is projective. Thus a Dedekind domain is noetherian because every ideal is projective (hence every ideal is finitely generated). See [**12**], [**25**], and [**31**] for more details on Dedekind domains.

For an invertible ideal I of a commutative domain R, let

$$I^{-2} = I^{-1}I^{-1}, \ I^{-3} = I^{-1}I^{-1}I^{-1},$$

and so on.

Let R be a commutative domain with the field of fractions F. It is well-known that if R is Prüfer, then any intermediate domain between R and F is also Prüfer.

Proposition 3.7 ([**21**, Lemma 2.8 and Lemma 2.9]). *Let R be a commutative domain. Then we have the following.*

(i) For invertible ideals $I_1, \dots, I_n$ of R, $(I_1 \cdots I_n)^{-1} = I_n^{-1} \cdots I_1^{-1}$.

(ii) For an invertible ideal I of R, $(I^\ell)^{-1} = I^{-\ell}$ for any non-negative integer ℓ.

(iii) If I is an invertible ideal of R, then $T(I) = \sum_{\ell\geq 0} I^{-\ell}$ and

$$T(I) = R[q_1, q_2, \dots, q_n],$$

where $1 = \sum_{i=1}^n r_i q_i$ for some $r_i \in I$ and $q_i \in I^{-1}$ with $1 \leq i \leq n$. Additionally, if R is a Dedekind (resp., Prüfer) domain, then $T(I)$ is a Dedekind (resp., Prüfer) domain.

Theorem 3.8 ([**21**, Theorem 2.13]). *Let R be a Dedekind domain. Assume that M is an R-module such that $I := Ann_R(M) \neq 0$, and $\{K_1, K_2, \dots, K_m\}$ is a finite set of nonzero finitely generated submodules of F_R, where F is the field of fractions of R. Then the following are equivalent.*

(i) $M_R \oplus (\oplus_{i=1}^m K_i)_R$ has a Baer hull.

(ii) $M_R \oplus (\oplus_{i=1}^m K_i)_R$ *has a Baer essential extension.*
(iii) M_R *is semisimple.*

In this case, $\mathbf{B}(M_R \oplus (\oplus_{i=1}^m K_i)_R) = M_R \oplus (\oplus_{i=1}^m K_i T(I))_R$. *Furthermore,* $T(I) = R[q_1, q_2, \ldots, q_n]$, *where* $1 = \sum_{k=1}^n a_k q_k$ *with* $a_k \in I$ *and* $q_k \in I^{-1}$, $1 \le k \le n$ *for some positive integer* n.

Let R be a commutative domain with the field of fractions F. An R-submodule K of F_R is called a fractional ideal of R if there exists $0 \ne r \in R$ such that $rK \subseteq R$. Every finitely generated R-submodule of F_R is a fractional ideal of R. Further, if R is noetherian, then a submodule K of F_R is fractional ideal of R if and only if K is finitely generated as an R-module.

The next theorem details the structure of finitely generated modules over a Dedekind domain.

THEOREM 3.9 (see [**31**, Theorem 6.16, p.177]). *Let R be a Dedekind domain and N a finitely generated R-module. Then there exist positive integers $n_1, \ldots, n_k$ (k is a non-negative integer), nonzero maximal ideals $P_1, \ldots, P_k$, and nonzero fractional ideals $K_1, \ldots, K_m$ (m is a non-negative integer) of R such that $N \cong (\oplus_{i=1}^k R/P_i^{n_i}) \oplus (\oplus_{j=1}^m K_j)$ as R-modules.*

We remark that if N is a torsion-free finitely generated module over a Dedekind domain R, then Theorem 3.9 yields that N is a projective module over R. By Theorem 3.9, the following theorem is a restatement of Theorem 3.8 in terms of a finitely generated module over a Dedekind domain.

THEOREM 3.10 ([**21**, Theorem 2.18]). *Let R be a Dedekind domain and N be a finitely generated R-module. Then the following are equivalent.*
(i) N *has a Baer hull.*
(ii) N *has a Baer essential extension.*
(iii) $t(N)$ *is semisimple.*

In this case, $\mathbf{B}(N) \cong t(N) \oplus [N/t(N)]T(Ann_R(t(N)))$.

EXAMPLE 3.11 ([**21**, Example 3.7(ii)]). We note that in the chain of $\mathbb{Z}$-submodules

$$\mathbb{Z}_p \le \mathbb{Z}_{p^2} \le \cdots \le \mathbb{Z}_{p^\infty}$$

of $\mathbb{Z}_{p^\infty}$ (p a prime integer), $\mathbb{Z}_p$ is the Baer hull of itself and $\mathbb{Z}_{p^\infty}$ is the injective hull of each of the modules in the chain. However, $\mathbb{Z}_{p^n}$ ($n > 1$) has no Baer hull by Theorem 3.8 (or Theorem 3.10). Note also that $\mathbb{Z}_{p^\infty}$ has no Baer hull.

Note that $\mathbb{Z}_p \oplus \mathbb{Z}$ is not a Baer $\mathbb{Z}$-module by Proposition 3.2. Our next example exhibits an explicit construction of the Baer hull of the module $\mathbb{Z}_p \oplus \mathbb{Z}$.

EXAMPLE 3.12. Let $N = \mathbb{Z}_p \oplus \mathbb{Z}$ (p a prime integer) as a $\mathbb{Z}$-module. Because $t(N) = \mathbb{Z}_p$ is simple, N has a Baer hull by Theorem 3.8 (or Theorem 3.10). We put $M = \mathbb{Z}_p$. Then $I := \mathrm{Ann}_{\mathbb{Z}}(M) = p\mathbb{Z}$.

First, note that $1/p \in (p\mathbb{Z})^{-1}$ and so $(1/p)\mathbb{Z} \subseteq (p\mathbb{Z})^{-1}$. Conversely, let $m/n \in (p\mathbb{Z})^{-1}$, where integers m and n are relatively prime, and n is positive. Because $(m/n)p\mathbb{Z} \subseteq \mathbb{Z}$, $(mp)/n := a \in \mathbb{Z}$, and therefore $mp = na$. Furthermore, since m and n are relatively prime, n is a divisor of p. But since n is a positive integer, $n = 1$ or $n = p$. Hence

$$m/n = m \quad \text{or} \quad m/n = m/p.$$

So $m/n \in (1/p)\mathbb{Z}$. Thus $(p\mathbb{Z})^{-1} \subseteq (1/p)\mathbb{Z}$. Consequently, $(p\mathbb{Z})^{-1} = (1/p)\mathbb{Z}$. By Proposition 3.7,

$$\begin{aligned}(p\mathbb{Z})^{-2} &= ((p\mathbb{Z})^2)^{-1} = ((p\mathbb{Z})(p\mathbb{Z}))^{-1} \\ &= (p\mathbb{Z})^{-1}(p\mathbb{Z})^{-1} = (1/p)\mathbb{Z}\,(1/p)\mathbb{Z} \\ &= (1/p^2)\mathbb{Z}.\end{aligned}$$

In general, $(p\mathbb{Z})^{-k} = (1/p^k)\mathbb{Z}$ for any positive integer k. Hence

$$T(I) = \sum_{\ell\geq 0} I^{-\ell} = \sum_{\ell \geq 0}(p\mathbb{Z})^{-\ell} = \mathbb{Z} + (1/p)\mathbb{Z} + (1/p^2)\mathbb{Z} + \cdots = \mathbb{Z}[1/p]$$

since $I^0 = (p\mathbb{Z})^0 = \mathbb{Z}$. Therefore

$$\mathbf{B}(N) = \mathbb{Z}_p \oplus \mathbb{Z}T(I) = \mathbb{Z}_p \oplus \mathbb{Z}[1/p]$$

by Theorem 3.8 or Theorem 3.10.

The following example also illustrates Theorem 3.8, Theorem 3.10, and presents a more general construction of Baer hulls.

EXAMPLE 3.13. (i) Let $p_1, p_2, \ldots, p_n$ be prime integers (which may not be distinct) and $p_1, p_2, \ldots, p_s$ be all the distinct prime integers in $\{p_1, p_2, \ldots, p_n\}$. Now put $a = p_1 p_2 \cdots p_s$. Consider

$$N = \mathbb{Z}_{p_1} \oplus \mathbb{Z}_{p_2} \oplus \cdots \oplus \mathbb{Z}_{p_n} \oplus \mathbb{Z}^{(m)},$$

where m is a non-negative integer. From Theorem 3.8 (or Theorem 3.10), N has a Baer hull since $t(N)$ is semisimple. Let $M = t(N)$. Then $I := \mathrm{Ann}(M) = a\mathbb{Z}$.

We compute $T(I)$. First, $I^0 = \mathbb{Z}$. Next for any positive integer k, we have that

$$\begin{aligned}I^{-k} &= (a\mathbb{Z})^{-k} = ((p_1 \cdots p_s)\mathbb{Z})^{-k} = ([(p_1\mathbb{Z}) \cdots (p_s\mathbb{Z})]^k)^{-1} \\ &= [(p_1\mathbb{Z})^k \cdots (p_s\mathbb{Z})^k]^{-1} = [(p_1^k\mathbb{Z}) \cdots (p_s^k\mathbb{Z})]^{-1} \\ &= (p_1^k\mathbb{Z})^{-1} \cdots (p_s^k\mathbb{Z})^{-1} = (1/p_1^k)\mathbb{Z} \cdots (1/p_s^k)\mathbb{Z} \\ &= (1/a^k)\mathbb{Z}\end{aligned}$$

by Proposition 3.7 and the argument used in Example 3.12. Thus

$$T(I) = \mathbb{Z} + (1/a)\mathbb{Z} + (1/a^2)\mathbb{Z} + \cdots = \mathbb{Z}[1/a].$$

Therefore by Theorem 3.8 (or Theorem 3.10),

$$\mathbf{B}(N) = \mathbb{Z}_{p_1} \oplus \mathbb{Z}_{p_2} \oplus \cdots \oplus \mathbb{Z}_{p_n} \oplus \mathbb{Z}^{(m)}T(I) = \mathbb{Z}_{p_1} \oplus \mathbb{Z}_{p_2} \oplus \cdots \oplus \mathbb{Z}_{p_n} \oplus \mathbb{Z}[1/a]^{(m)}.$$

(ii) Let $V = \mathbb{Z}_{{p_1}^{n_1}} \oplus \mathbb{Z}_{{p_2}^{n_2}} \oplus \cdots \oplus \mathbb{Z}_{{p_k}^{n_k}} \oplus \mathbb{Z}^{(m)}$, where k is a positive integer, $p_1, p_2, \ldots, p_k$ are prime integers (not necessarily to be distinct), $n_1, n_2, \ldots, n_k$ are positive integers, and m is a non-negative integer. Then from Theorem 3.8 (or Theorem 3.10), $\mathbf{B}(V)$ exists if and only if $n_1 = \cdots = n_k = 1$. In this case, by (i)

$$\mathbf{B}(V) = \mathbb{Z}_{p_1} \oplus \mathbb{Z}_{p_2} \oplus \cdots \oplus \mathbb{Z}_{p_k} \oplus \mathbb{Z}[1/a]^{(m)}$$

with $a = p_1 \cdots p_s$ and $p_1, \ldots, p_s$ all the distinct prime integers in $\{p_1, \ldots, p_k\}$.

Recall that a ring R is called semiprimary if $R/J(R)$ is artinian and $J(R)$ is nilpotent, where $J(R)$ is the Jacobson radical of R.

THEOREM 3.14 ([**32**, Theorem 2] and [**28**, Theorem 3.3]). *Let R be a ring. Then the following are equivalent.*

(i) R is semiprimary right (left) hereditary.

(ii) $CFM_\Lambda(R)$ is a Baer ring for any nonempty set Λ.

From Theorem 3.8, one may expect that the assumption "$\{K_1, K_2, \cdots, K_m\}$ is a finite set of finitely generated nonzero R-submodules of F_R" can be extended to "$\{K_i \mid i \in \Lambda\}$ is a set (*not necessarily finite*) of finitely generated nonzero R-submodules of F_R". However, the following example shows that this is impossible.

EXAMPLE 3.15 ([**21**, Example 3.6]). Let $M = \oplus_{i=1}^n \mathbb{Z}_{p_i}$, where n is a positive integer, and all p_i are prime integers. Let $p_1, p_2, \ldots, p_s$ be all the distinct prime integers in $\{p_1, p_2, \ldots, p_n\}$. We take $a = p_1 p_2 \cdots p_s$ as in Example 3.13.

Since $\mathbb{Z}[1/a]$ is not a field, $\mathbb{Z}[1/a]$ is not semiprimary. By Theorem 3.14, there exists a nonempty set Λ such that $\text{CFM}_\Lambda(\mathbb{Z}[1/a])$ is not a Baer ring.

Furthermore, we have the following.

(i) $M \oplus \mathbb{Z}[1/a]^{(\Lambda)}$ is not a Baer $\mathbb{Z}$-module.

(ii) Let $N = M \oplus \mathbb{Z}^{(\Lambda)}$. Then $N/t(N)$ is not finitely generated.

(iii) $N = M \oplus \mathbb{Z}^{(\Lambda)}$ has *no Baer hull* as a $\mathbb{Z}$-module.

(i) Note that by Theorem 2.30(ii), $\mathbb{Z}[1/a]^{(\Lambda)}$ is quasi-retractable as a $\mathbb{Z}$-module, and

$$\text{End}_\mathbb{Z}(\mathbb{Z}[1/a]^{(\Lambda)}) = \text{End}_{\mathbb{Z}[1/a]}(\mathbb{Z}[1/a]^{(\Lambda)}) = \text{CFM}_\Lambda(\mathbb{Z}[1/a]).$$

Since $\text{End}_\mathbb{Z}(\mathbb{Z}[1/a]^{(\Lambda)}) = \text{CFM}_\Lambda(\mathbb{Z}[1/a])$ is not a Baer ring, $\mathbb{Z}[1/a]^{(\Lambda)}$ is not a Baer $\mathbb{Z}$-module by Theorem 2.27.

(ii) We prove that Λ is an infinite set. Assume on the contrary that Λ is a finite set, say $|\Lambda| = m$, where m is a positive integer. Note that $\mathbb{Z}[1/a]$ is a Prüfer domain. So $\text{End}_\mathbb{Z}(\mathbb{Z}[1/a]^{(\Lambda)}) = \text{Mat}_m(\mathbb{Z}[1/a])$ is a Baer ring by Theorem 2.9. Thus we get a contradiction. Therefore Λ is infinite, so $N/t(N) \cong \mathbb{Z}^{(\Lambda)}$ is not finitely generated as a $\mathbb{Z}$-module.

(iii) Since $\mathbb{Z}[1/a]^{(\Lambda)}$ is not a Baer $\mathbb{Z}$-module by the proof of (i), Theorem 3.22 yields that N has no Baer hull. However, we give the proof in detail for illustrating previously mentioned results on Baer hulls.

For each $\alpha \in \Lambda$, let $W_\alpha = \oplus_{i\in\Lambda} U_i$, where $U_\alpha = \mathbb{Z}[1/a]$ and $U_i = \mathbb{Q}$ for $i \neq \alpha$. First we claim that W_α is a Baer $\mathbb{Z}$-module. For this, note that $\mathbb{Z}[1/a]$ is a nonsingular extending $\mathbb{Z}$-module and $\oplus_{i\in\Lambda\setminus\{\alpha\}} U_i = \mathbb{Q}^{(\Lambda\setminus\{\alpha\})}$ is a nonsingular injective $\mathbb{Z}$-module. By Theorem 2.40, W_α is a Baer $\mathbb{Z}$-module for each $\alpha \in \Lambda$.

Assume on the contrary that $M \oplus \mathbb{Z}^{(\Lambda)}$ has a Baer hull, say V. Take

$$f \in \text{Hom}_\mathbb{Z}(\mathbb{Z}[1/a], M).$$

Then $f(1) = af(1/a) = 0$, $f(1/a) = af(1/a^2) = 0$, and so on. Thus $f(1/a^k) = 0$ for any integer k such that $k \geq 0$. Therefore $f = 0$, hence

$$\text{Hom}_\mathbb{Z}(\mathbb{Z}[1/a], M) = 0.$$

Next let $g \in \text{Hom}(\mathbb{Q}, M)$. Then for any nonzero integer n, $f(1/n) = af(1/na) = 0$. Thus for any $q \in \mathbb{Q}$ with $q = m/n$, $f(q) = f(m/n) = mf(1/n) = 0$. Hence $g = 0$, so $\text{Hom}_\mathbb{Z}(\mathbb{Q}, M) = 0$. Therefore

$$\text{Hom}_\mathbb{Z}(\mathbb{Q}^{(\Lambda\setminus\{\alpha\})}, M) = 0$$

for each $\alpha \in \Lambda$. Consequently, $\mathrm{Hom}_{\mathbb{Z}}(W_\alpha, M) = 0$. So $M \oplus W_\alpha$ is a Baer module for each $\alpha \in \Lambda$ by Theorem 3.6. Thus

$$V \subseteq \cap_{\alpha\in\Lambda}(M \oplus W_\alpha) = M \oplus (\cap_{\alpha\in\Lambda} W_\alpha) = M \oplus \mathbb{Z}[1/a]^{(\Lambda)}$$

because $\cap_{\alpha\in\Lambda} W_\alpha = \mathbb{Z}[1/a]^{(\Lambda)}$.

We note that V is a Baer module and $M \oplus \mathbb{Z}^{(\Lambda)} \le V \le E(M \oplus \mathbb{Z}^{(\Lambda)})$. Hence from Theorem 3.6, $V = M \oplus W$ for some W such that

$$\mathbb{Z}^{(\Lambda)} \le W \le E(\mathbb{Z}^{(\Lambda)}) = \mathbb{Q}^{(\Lambda)},\ W \text{ is a Baer module, and } \mathrm{Hom}_{\mathbb{Z}}(W, M) = 0.$$

We put $I = \mathrm{Ann}_{\mathbb{Z}}(M) = a\mathbb{Z}$. Since $M = \oplus_{i=1}^n \mathbb{Z}_{p_i}$, it follows that $\mathrm{Hom}_{\mathbb{Z}}(W, \mathbb{Z}_{p_i}) = 0$ for each i, $1 \le i \le n$.

Put $P = p_1\mathbb{Z}$. We claim that $WP = W$. Indeed, from the exact sequence

$$0 \to WP \to W \to W/WP \to 0,$$

we obtain the following exact sequence

$$0 \to \mathrm{Hom}_{\mathbb{Z}}(W/WP, \mathbb{Z}/P) \to \mathrm{Hom}_{\mathbb{Z}}(W, \mathbb{Z}/P) \to \mathrm{Hom}_{\mathbb{Z}}(WP, \mathbb{Z}/P)$$

because $\mathrm{Hom}_{\mathbb{Z}}(-, \mathbb{Z}/P)$ is a left exact contravariant functor.

As $\mathrm{Hom}_{\mathbb{Z}}(W, \mathbb{Z}/P) = \mathrm{Hom}_{\mathbb{Z}}(W, \mathbb{Z}_{p_1}) = 0$, from the preceding exact sequence we have that $\mathrm{Hom}_{\mathbb{Z}}(W/WP, \mathbb{Z}/P) = 0$. We notice that W/WP is a $\mathbb{Z}/P$-module, which is induced from the $\mathbb{Z}$-module structure of W/WP.

Further, $\mathrm{Hom}_{\mathbb{Z}/P}(W/WP, \mathbb{Z}/P) = \mathrm{Hom}_{\mathbb{Z}}(W/WP, \mathbb{Z}/P) = 0$. Since $\mathbb{Z}/P$ is a field, W/WP is a vector space over $\mathbb{Z}/P$. Therefore $W/WP = 0$, and thus $W = WP$. So $W(p_1\mathbb{Z}) = W$. Similarly $W(p_i\mathbb{Z}) = W$ for each $i, 1 \le i \le m$.

Now $I = a\mathbb{Z} = p_1p_2\cdots p_s\mathbb{Z} = (p_1\mathbb{Z})(p_2\mathbb{Z})\cdots(p_s\mathbb{Z})$, and so $WI = W$. Put $T(I) = \sum_{\ell\ge 0} I^{-\ell}$, the Nagata transform of I. Then

$$T(I) = \mathbb{Z}[1/a]$$

as in Example 3.13.

We claim that $T(I)^{(\Lambda)} = (\mathbb{Z}T(I))^{(\Lambda)} \subseteq W$. For this, first consider the case when $|\Lambda| = 2$. We let $J = I^\ell$, where ℓ is a non-negative integer. Since $WI = W$, $WI^\ell = W$ and so $WJ = W$. We now take $\beta \in \mathbb{Z}$. Then

$$(\beta, 0) \in \mathbb{Z} \oplus \mathbb{Z} \subseteq W = WJ.$$

Hence there exists a positive integer n such that $(\beta, 0) = \sum_{i=1}^n w_i a_i$, where $w_i \in W$ and $a_i \in J$ for i, $1 \le i \le n$.

Put $w_i = (x_i, y_i) \in E(\mathbb{Z}_{\mathbb{Z}})^{(2)} = \mathbb{Q} \oplus \mathbb{Q}$ for i, $1 \le i \le n$. Then

$$(\beta, 0) = \left(\sum_{i=1}^n x_i a_i, \sum_{i=1}^n y_i a_i\right).$$

So $\beta = \sum_{i=1}^n x_i a_i$ and $0 = \sum_{i=1}^n y_i a_i$. Take any $q \in J^{-1}$. Then

$$\beta q = \sum_{i=1}^n x_i q a_i \quad\text{and}\quad 0 = \sum_{i=1}^n y_i q a_i.$$

Since each $a_i \in J$ and $qJ \subseteq \mathbb{Z}$, each $qa_i \in \mathbb{Z}$. So

$$(\beta q, 0) = \left(\sum_{i=1}^n x_i q a_i, \sum_{i=1}^n y_i q a_i\right) = \sum_{i=1}^n (x_i, y_i) q a_i = \sum_{i=1}^n w_i q a_i.$$

Hence $(\beta q, 0) \in \sum_{i=1}^{n} w_i\mathbb{Z} \subseteq W$. Therefore $(\mathbb{Z}J^{-1}, 0) \subseteq W$. By Proposition 3.7, $J^{-1} = I^{-\ell}$. Thus $(\mathbb{Z}I^{-\ell}, 0) \subseteq W$ for all non-negative integers ℓ. Hence it follows that$(\mathbb{Z}T(I), 0) \subseteq W$, and thus $(T(I), 0) \subseteq W$.

Similarly, $(0, T(I)) \subseteq W$. So $T(I) \oplus T(I) \subseteq W$. In general, by applying the preceding method, $T(I)^{(\Lambda)} \leq W$ as $\mathbb{Z}^{(\Lambda)} \leq W$. So $\mathbb{Z}[1/a]^{(\Lambda)} \subseteq W$. Therefore $M \oplus \mathbb{Z}[1/a]^{(\Lambda)} \leq M \oplus W = V$.

Consequently, $V = M \oplus \mathbb{Z}[1/a]^{(\Lambda)}$, and thus $M \oplus \mathbb{Z}[1/a]^{(\Lambda)}$ is a Baer module, which is a contradiction to (i). Hence $M \oplus \mathbb{Z}^{(\Lambda)}$ has no Baer hull.

We recall that **Ric**$(-)$ denotes the Rickart hull of a module.

THEOREM 3.16 ([**15**, Lemma 2.11 and Theorem 3.3]). *Let R be a Dedekind domain, M an R-module with $I := Ann_R(M) \neq 0$, and let $\{K_i \mid i \in \Lambda\}$ be a set of nonzero finitely generated submodules of F_R, where F is the field of fractions of R. Then the following are equivalent.*

(i) $M_R \oplus (\oplus_{i\in\Lambda} K_i)_R$ has a Rickart hull.

(ii) $M_R \oplus (\oplus_{i\in\Lambda} K_i)_R$ has a Rickart essential extension.

(iii) M_R is semisimple.

In this case, $\mathbf{Ric}(M_R \oplus (\oplus_{i\in\Lambda} K_i)_R) = M_R \oplus (\oplus_{i\in\Lambda} K_i T(I))_R$. *Furthermore, $T(I) = R[q_1, q_2, \ldots, q_n]$, where $1 = \sum_{k=1}^{n} a_k q_k$ with $a_k \in I$ and $q_k \in I^{-1}$, $1 \leq k \leq n$ for some positive integer n.*

EXAMPLE 3.17. Let $M = \oplus_{i=1}^{n} \mathbb{Z}_{p_i}$, $a = p_1 p_2 \cdots p_s$, and Λ be as in Example 3.15. Then $I := \mathrm{Ann}_R(M) = aR$, and so $T(I) = \mathbb{Z}[1/a]$. Therefore $M \oplus \mathbb{Z}[1/a]^{(\Lambda)}$ is the Rickart hull of $M \oplus \mathbb{Z}^{(\Lambda)}$ by Theorem 3.16. However, we note that $M \oplus \mathbb{Z}^{(\Lambda)}$ has no Baer hull by Example 3.15.

See [**29**, Theorem 4.17, p.121] for the proof of Lemma 3.18.

LEMMA 3.18. *Let R be a right hereditary ring. Then every submodule of a free right R-module is isomorphic to a direct sum of right ideals of R (as R-modules).*

Let N be a module over a Dedekind domain R such that $N/t(N)$ is projective. Then as $N/t(N)$ is projective, $N \cong t(N) \oplus N/t(N)$. By Lemma 3.18, it follows that $N/t(N) \cong \oplus_{i\in\Lambda} K_i$, where $\{K_i \mid i \in \Lambda\}$ is a set of right ideals of R. Since R is Dedekind, R is noetherian and hence each K_i is finitely generated as an R-module.

We note that the next result is a restatement of Theorem 3.16. If V is a finitely generated module over a Dedekind domain R, then $V/t(V)$ is projective and $\mathrm{Ann}_R(t(V)) \neq 0$ from 3.9. Thus it is of interest to compare it with Theorem 3.10.

THEOREM 3.19 ([**15**, Lemma 2.11 and Theorem 4.2]). *Let R be a Dedekind domain. Assume that N is an R-module with $N/t(N)$ projective and $Ann_R(t(N)) \neq 0$. Then the following are equivalent.*

(i) N has a Rickart hull.

(ii) N has a Rickart essential extension.

(iii) $t(N)$ is semisimple.

In this case, $\mathbf{Ric}(N) \cong t(N) \oplus (N/t(N)T(Ann_R(t(N))))$.

The Rickart and Baer hulls sometimes coincide as shown by our next a few results.

COROLLARY 3.20 (see [**15**, Corollary 4.5]). *Let R be a Dedekind domain, M an R-module with $I := Ann_R(M) \neq 0$, and let $\{K_1, K_2, \ldots, K_m\}$ be a finite set of nonzero finitely generated submodules of F_R, where F is the field of fractions of R. We put $N = M_R \oplus (\oplus_{i=1}^m K_i)_R$. Then the following are equivalent.*

(i) N has a Rickart hull.

(ii) N has a Baer hull.

(iii) N has a Rickart essential extension.

(iv) N has a Baer essential extension.

(v) M_R is semisimple.

In this case, $\mathbf{Ric}(N) = \mathbf{B}(N) = M \oplus (\oplus_{i=1}^m K_i T(I))$. Furthermore, $T(I) = R[q_1, q_2, \ldots, q_n]$, where $1 = \sum_{k=1}^n a_k q_k$ with $a_k \in I$ and $q_k \in I^{-1}$, $1 \leq k \leq n$ for some positive integer n.

COROLLARY 3.21. *Let R be a Dedekind domain and N be a finitely generated R-module. Then the following are equivalent.*

(i) N has a Rickart hull.

(ii) N has a Baer hull.

(iii) N has a Rickart essential extension.

(iv) N has a Baer essential extension.

(v) $t(N)$ is semisimple.

In this case, $\mathbf{Ric}(N) = \mathbf{B}(N) \cong t(N) \oplus [N/t(N)]T(Ann_R(t(N)))$.

Motivated by Example 3.15, we have the following result, which is related to the possibility of extensions of hypotheses in Theorem 3.8 and Theorem 3.16.

THEOREM 3.22 ([**15**, Theorem 3.6]). *Let R be a Dedekind domain with F the field of fractions, M an R-module with $I := Ann_R(M) \neq 0$, and let $\{K_i \mid i \in \Lambda\}$ be a set of nonzero submodules of F_R. Put $N = M \oplus (\oplus_{i\in\Lambda} K_i)$. Then the following are equivalent.*

(i) N has a Baer (resp., Rickart) hull.

(ii) M is semisimple and $\oplus_{i\in\Lambda} K_i T(I)$ is a Baer (resp., Ricakrt) module.

In this case, $\mathbf{B}(N) = M \oplus (\oplus_{i\in\Lambda} K_i T(I))$ (resp., $\mathbf{Ric}(N) = M \oplus (\oplus_{i\in\Lambda} K_i T(I))$).

From Example 3.15, we may expect that the assumption "$\{K_1, K_2, \cdots, K_m\}$ is a finite set of finitely generated nonzero R-submodules of F_R" in Theorem 3.8 can be extended to "$\{K_1, K_2, \ldots, K_m\}$ is a finite set of of nonzero R-submodules (*not necessarily finitely generated*) of F_R".

Furthermore, one may also expect that the assumption "$\{K_i \mid i \in \Lambda\}$ is a set of nonzero finitely generated submodules of F_R" in Theorem 3.16 can be extended to "$\{K_i \mid i \in \Lambda\}$ is a set of nonzero submodules of F_R".

However, the following example eliminates each of these two expectations.

EXAMPLE 3.23 (see [**15**, Example 3.5]). Let $W = \mathbb{Z}[1/6] \oplus \mathbb{Z}[1/2]$. We show that W is not a Rickart module. For this, take $f \in \mathrm{Hom}_{\mathbb{Z}}(\mathbb{Z}[1/6], \mathbb{Z}[1/2])$. Then for any non-negative integer k, $f(1) = 6^k f(1/6^k)$ and hence $f(1/6^k) = (1/6^k)f(1)$.

Say $f(1) = a/2^k$, where $a \in \mathbb{Z}$ and k is a non-negative integer. Then for any positive integer n, we have that

$$f(1/6^n) = (1/6^n)f(1) = (1/6^n)(a/2^k) = a/(6^n 2^k) \in \mathbb{Z}[1/2].$$

Thus there exist a non-negative integer m and an integer b such that

$$a/(6^n 2^k) = b/2^m, \quad \text{hence} \quad 6^n 2^k b = 2^m a.$$

So $3^n 2^n 2^k b = 2^m a$. Thus 3^n divides a for each positive integer n. Hence $a = 0$, so $f(1) = 0$. Therefore $f = 0$, and thus $\mathrm{Hom}_{\mathbb{Z}}(\mathbb{Z}[1/6], \mathbb{Z}[1/2]) = 0$.

Next $\mathrm{Hom}_{\mathbb{Z}}(\mathbb{Z}[1/2], \mathbb{Z}[1/6]) \cong \mathbb{Z}[1/6]$ as abelian groups. Indeed, define

$$\theta : \mathrm{Hom}_{\mathbb{Z}}(\mathbb{Z}[1/2], \mathbb{Z}[1/6]) \to \mathbb{Z}[1/6]$$

by $\theta(g) = g(1)$ for $g \in \mathrm{Hom}_{\mathbb{Z}}(\mathbb{Z}[1/2], \mathbb{Z}[1/6])$. Then θ is a one-to-one additive group homomorphism. Now for $\beta \in \mathbb{Z}[1/6]$, let $g(t/2^k) = \beta \cdot (t/2^k)$ for $t/2^k \in \mathbb{Z}[1/2]$ with $t \in \mathbb{Z}$. Then $g(1) = \beta$. Therefore θ is onto, so

$$\mathrm{Hom}_{\mathbb{Z}}(\mathbb{Z}[1/2], \mathbb{Z}[1/6]) \cong \mathbb{Z}[1/6]$$

as additive abelian groups. Furthermore, we note that $\mathrm{End}_{\mathbb{Z}}(\mathbb{Z}[1/6]) \cong \mathbb{Z}[1/6]$ and $\mathrm{End}_{\mathbb{Z}}(\mathbb{Z}[1/2]) \cong \mathbb{Z}[1/2]$. Hence

$$\mathrm{End}_{\mathbb{Z}}(W) = \begin{bmatrix} \mathrm{End}_{\mathbb{Z}}(\mathbb{Z}[1/6]) & \mathrm{Hom}_{\mathbb{Z}}(\mathbb{Z}[1/2], \mathbb{Z}[1/6]) \\ \mathrm{Hom}_{\mathbb{Z}}(\mathbb{Z}[1/6], \mathbb{Z}[1/2]) & \mathrm{End}_{\mathbb{Z}}(\mathbb{Z}[1/2]) \end{bmatrix} \cong \begin{bmatrix} \mathbb{Z}[1/6] & \mathbb{Z}[1/6] \\ 0 & \mathbb{Z}[1/2] \end{bmatrix}.$$

Take

$$\varphi = \begin{bmatrix} 5 & -1/3 \\ 0 & 0 \end{bmatrix} \in \mathrm{End}_{\mathbb{Z}}(W) \text{ and note that } \begin{bmatrix} 1 \\ 15 \end{bmatrix} \in \mathrm{Ker}(\varphi).$$

Moreover, we notice $\mathrm{Ker}(\varphi) \neq W$. So if there exists $e^2 = e \in \mathrm{End}_{\mathbb{Z}}(W)$ such that $\mathrm{Ker}(\varphi) = eW$, then there exists $\beta \in \mathbb{Z}[1/6]$ and

$$e = \begin{bmatrix} 0 & \beta \\ 0 & 1 \end{bmatrix}, \text{ hence } \begin{bmatrix} 1 \\ 15 \end{bmatrix} = \begin{bmatrix} 0 & \beta \\ 0 & 1 \end{bmatrix} \begin{bmatrix} u \\ v \end{bmatrix} \text{ with } u \in \mathbb{Z}[1/6] \text{ and } v \in \mathbb{Z}[1/2].$$

Thus $\beta v = 1$ and $v = 15$, so $\beta = 1/15$. Therefore

$$1/15 \in \mathbb{Z}[1/6].$$

Hence $1/15 = a/6^b$ for an integer a and a non-negative integer b. So

$$15a = 6^b.$$

If $b = 0$, then $15a = 1$ from $15a = 6^b$, so it is absurd. If $b = 1$, then from $15a = 6^b$ we have that $5a = 2$. So this is also absurd. Finally, if $b \geq 2$, then from $15a = 6^b$, it follows that $5a = 2^b 3^{b-1}$. So 5 divides either 2 or 3, which is a contradiction.

Consequently, $1/15 \notin \mathbb{Z}[1/6]$, so there is no $e^2 = e \in \mathrm{End}(W)$ such that $\mathrm{Ker}(\varphi) = eW$. Hence W is not a Rickart module. So W is not a Baer module. Therefore by Theorem 3.22, W has no Baer hull and W has no Rickart hull.

In Proposition 3.24 and Example 3.25, we consider the isomorphism problem for Baer (resp., Rickart) hulls that is: Let N_1 and N_2 be modules with Baer (resp., Rickart) hulls $\mathbf{B}(N_1)$ and $\mathbf{B}(N_2)$ (resp., $\mathbf{Ric}(N_1)$ and $\mathbf{Ric}(N_2)$). Then it is natural to ask: Is it true that $N_1 \cong N_2$ if and only if $\mathbf{B}(N_1) \cong \mathbf{B}(N_2)$ (resp., $\mathbf{Ric}(N_1) \cong \mathbf{Ric}(N_2)$)? Our next result and Example 3.25 show that this works only in one direction.

PROPOSITION 3.24. *Let N_1 and N_2 be isomorphic modules.*

(i) If N_1 has a Baer hull $\mathbf{B}(N_1)$, then N_2 has a Baer hull $\mathbf{B}(N_2)$, and $\mathbf{B}(N_1) \cong \mathbf{B}(N_2)$ as modules.

(ii) If N_1 has a Rickart hull $\mathbf{Ric}(N_1)$, then N_2 has a Rickart hull $\mathbf{Ric}(N_2)$, and $\mathbf{Ric}(N_1) \cong \mathbf{Ric}(N_2)$ as modules.

The next example shows that the converse of Proposition 3.24 does not hold true. Indeed, there exist modules N_1 and N_2 such that $\mathbf{B}(N_1) = \mathbf{B}(N_2)$ (also $\mathbf{Ric}(N_1) = \mathbf{Ric}(N_2)$), but $N_1 \not\cong N_2$. Thus the isomorphism problem does not fully hold for the case of Baer hulls as well as Rickart hulls.

EXAMPLE 3.25 ([**21**, Example 3.9], [**22**, Example 4.14], and [**15**, Example 4.8]). Let $N_1 = \mathbb{Z}_2 \oplus \mathbb{Z}_3 \oplus \mathbb{Z}$. Then $\mathbb{Z}_2 \oplus \mathbb{Z}_3 \oplus \mathbb{Z}[1/6]$ is the Baer hull of N_1 as a $\mathbb{Z}$-module by Theorem 3.8.

Next, let $N_2 = \mathbb{Z}_2 \oplus \mathbb{Z}_3 \oplus \mathbb{Z}[1/3]$. Say V is a Baer module such that $N_2 \leq V \leq E(N_2)$. From Theorem 3.6,

$$V = \mathbb{Z}_2 \oplus Z_3 \oplus W$$

for some Baer module W such that $\mathbb{Z}[1/3] \leq W \leq \mathbb{Q}$ and $\mathrm{Hom}_{\mathbb{Z}}(W, \mathbb{Z}_2 \oplus \mathbb{Z}_3) = 0$. Thus $\mathrm{Hom}_{\mathbb{Z}}(W, \mathbb{Z}_2) = 0$, and so $W = W(2\mathbb{Z})$ by the argument used illustrating Example 3.15. Therefore $W = 2W$. Hence $2^k W = W$ for any non-negative integer k. Therefore $1/2^k \in W$ for any positive integer k, and thus $\mathbb{Z}[1/2, 1/3] \leq W$. Hence we obtain

$$\mathbb{Z}_2 \oplus \mathbb{Z}_3 \oplus \mathbb{Z}[1/2, 1/3] = \mathbb{Z}_2 \oplus \mathbb{Z}_3 \oplus \mathbb{Z}[1/6] \leq V.$$

We note that $\mathbb{Z}_2 \oplus \mathbb{Z}_3 \oplus \mathbb{Z}[1/6]$ is the Baer hull of $\mathbb{Z}_2 \oplus \mathbb{Z}_3 \oplus \mathbb{Z}$ by Theorem 3.8. Therefore $\mathbb{Z}_2 \oplus \mathbb{Z}_3 \oplus \mathbb{Z}[1/6]$ is Baer as a $\mathbb{Z}$-module, and so $\mathbb{Z}_2 \oplus \mathbb{Z}_3 \oplus \mathbb{Z}[1/6]$ is the Baer hull of N_2. Consequently,

$$\mathbf{B}(N_1) = \mathbf{B}(N_2) = \mathbb{Z}_2 \oplus \mathbb{Z}_3 \oplus \mathbb{Z}[1/6].$$

However, N_1 is not isomorphic to N_2 as $\mathbb{Z}$-modules. Indeed, if $N_1 \cong N_2$, then it follows that $\mathbb{Z} \cong \mathbb{Z}[1/3]$ as $\mathbb{Z}$-modules. But this is impossible because $\mathbb{Z}[1/3]$ is not finitely generated as a $\mathbb{Z}$-module.

From Theorem 3.16, $\mathbf{Ric}(N_1) = \mathbb{Z}_2 \oplus \mathbb{Z}_3 \oplus \mathbb{Z}[1/6]$. By Theorem 3.6 and the preceding argument, $\mathbf{Ric}(N_2) = \mathbb{Z}_2 \oplus \mathbb{Z}_3 \oplus \mathbb{Z}[1/6]$. So $\mathbf{Ric}(N_1) = \mathbf{Ric}(N_2)$. But $N_1 \not\cong N_2$ as above.

For Baer (resp., Rickart) modules U and V such that $U \leq^{\mathrm{ess}} V$, one may expect that any intermediate module between U and V is Baer (resp., Rickart). However, the next example shows that this expectation does not hold true.

EXAMPLE 3.26 (see [**15**, Example 3.11]). Let $W = \mathbb{Z}[1/6] \oplus \mathbb{Z}[1/2]$ as a $\mathbb{Z}$-module. Then we have the following.

(i) From Example 3.23, neither $\mathbf{Ric}(W)$ nor $\mathbf{B}(W)$ exists.

(ii) Let n be a positive integer which is a multiple of 6. Then

$$\mathbb{Z} \oplus \mathbb{Z} \leq^{\mathrm{ess}} W \leq^{\mathrm{ess}} \mathbb{Z}[1/n] \oplus \mathbb{Z}[1/n]$$

as $\mathbb{Z}$-modules. In this case, both $\mathbb{Z} \oplus \mathbb{Z}$ and $\mathbb{Z}[1/n] \oplus \mathbb{Z}[1/n]$ are quasi-retractable from Theorem 2.30(i). Note that

$$\mathrm{End}_{\mathbb{Z}}(\mathbb{Z}^{(2)}) = \mathrm{Mat}_2(\mathbb{Z}) \quad \text{and} \quad \mathrm{End}_{\mathbb{Z}}(\mathbb{Z}[1/n]^{(2)}) = \mathrm{Mat}_2(\mathbb{Z}[1/n])$$

are Baer by Theorem 2.9. So $\mathbb{Z} \oplus \mathbb{Z}$ and $\mathbb{Z}[1/n] \oplus \mathbb{Z}[1/n]$ are Baer (hence Rickart) $\mathbb{Z}$-modules from Theorem 2.32. However, W is neither Baer nor Rickart as a $\mathbb{Z}$-module by (i).

As an application of these classes of module hulls, we discuss connections between a Baer (resp., Rickart) hull of a direct summand of a module N and a Baer (resp., Rickart) hull of N.

Let R be a Dedekind domain. Assume that L and N are R-modules such that $L \leq^{\oplus} N$. Assume that $\mathbf{B}(N)$ exists. Then we may raise the following natural question: Is $\mathbf{B}(L) \leq^{\oplus} \mathbf{B}(N)$? How about this same question for the case of $\mathbf{Ric}(N)$? The following example provides a negative answer to both of these questions.

EXAMPLE 3.27. Let $N = \mathbb{Z}_p \oplus \mathbb{Z}$ (p a prime integer) and $L = \mathbb{Z}$ as $\mathbb{Z}$-modules. Then $L \leq^{\oplus} N$ as $\mathbb{Z}$-modules. By Theorem 3.8 and Theorem 3.16,

$$\mathbf{B}(N) = \mathbf{Ric}(N) = \mathbb{Z}_p \oplus \mathbb{Z}[1/p].$$

Note $\mathbf{B}(L) = \mathbf{Ric}(L) = \mathbb{Z}$. It can be checked that

$$\mathbf{B}(L) \not\leq^{\oplus} \mathbf{B}(N) \text{ and } \mathbf{Ric}(L) \not\leq^{\oplus} \mathbf{Ric}(N).$$

In fact, assume on the contrary that $\mathbf{B}(L) \leq^{\oplus} \mathbf{B}(N)$. Then $\mathbb{Z} \leq^{\oplus} \mathbb{Z} \oplus \mathbb{Z}[1/p]$ as $\mathbb{Z}$-modules. Thus $\mathbb{Z}_p \oplus \mathbb{Z}[1/p] = \mathbb{Z} \oplus W$ for some $W \leq \mathbb{Z}_p \oplus \mathbb{Z}[1/p]$. By the modular law, $\mathbb{Z}[1/p] = \mathbb{Z} \oplus (\mathbb{Z}[1/p] \cap W)$. Since $\mathbb{Z}[1/p]$ is uniform as a $\mathbb{Z}$-module, $\mathbb{Z}[1/p] \cap W = 0$. Hence $\mathbb{Z}[1/p] = \mathbb{Z}$, which is a contradiction because $\mathbb{Z}[1/p]$ is not cyclic as a $\mathbb{Z}$-module. Therefore $\mathbf{B}(L) \not\leq^{\oplus} \mathbf{B}(N)$. Similarly, $\mathbf{Ric}(L) \not\leq^{\oplus} \mathbf{Ric}(N)$.

Suppose that U and V are R-modules such that $\mathbf{B}(U)$ (resp., $\mathbf{Ric}(U)$), $\mathbf{B}(V)$ (resp., $\mathbf{Ric}(V)$), and $\mathbf{B}(U \oplus V)$ (resp., $\mathbf{Ric}(U \oplus V)$) exist. The following example shows that $\mathbf{B}(U \oplus V) = \mathbf{B}(U) \oplus \mathbf{B}(V)$ (resp., $\mathbf{Ric}(U \oplus V) = \mathbf{Ric}(U) \oplus \mathbf{Ric}(V)$) does not hold true in general.

EXAMPLE 3.28 ([**21**, Example 3.10] and [**22**, Example 4.16]). Let $U = \mathbb{Z}_p$ (p a prime integer) and $V = \mathbb{Z}$ as $\mathbb{Z}$-modules. Then

$$\mathbf{B}(U) = \mathbf{Ric}(U) = \mathbb{Z}_p \text{ and } \mathbf{B}(V) = \mathbf{Ric}(V) = \mathbb{Z}.$$

Hence $\mathbf{B}(U) \oplus \mathbf{B}(V) = \mathbb{Z}_p \oplus \mathbb{Z}$. Note that $\mathbf{B}(U \oplus V) = \mathbb{Z}_p \oplus \mathbb{Z}[1/p]$ (see Theorem 3.8). Suppose that $\mathbf{B}(U \oplus V) \cong \mathbf{B}(U) \oplus \mathbf{B}(V)$. Then $\mathbb{Z}[1/p] \cong \mathbb{Z}$ as $\mathbb{Z}$-modules, a contradiction. Therefore

$$\mathbf{B}(U \oplus V) \not\cong \mathbf{B}(U) \oplus \mathbf{B}(V).$$

Similarly, $\mathbf{Ric}(U \oplus V) \not\cong \mathbf{Ric}(U) \oplus \mathbf{Ric}(V)$.

While there exist obvious disparities between Baer and Rickart properties of modules, the next theorem shows that the two notions coincide for the case when a module is finitely generated over a Dedekind domain. This result extends Proposition 3.2.

THEOREM 3.29 ([**15**, Theorem 4.10]). *Let R be a Dedekind domain and N be a finitely generated R-module. Then the following are equivalent.*

(i) N is Baer.

(ii) N is Rickart.

(iii) N is semisimple or torsion-free.

The next example shows that a commutative domain R needs to be a Dedekind domain for the previous result to hold.

EXAMPLE 3.30 (see [**15**, Example 4.15]). There exists a commutative domain R for which Theorem 3.29 does not hold. Let $R = \mathbb{Z}[x]$ and $N = (R \oplus R)_R$, which is a finitely generated R-module. Then:

(i) N is a torsion-free R-module.

(ii) N is not a Rickart module by Theorem 2.43(ii). Indeed, we note that $\mathrm{End}_R(N) = \mathrm{Mat}_2(R)$ is not a right Rickart ring as the domain R is not Prüfer (see Theorem 2.9).

(iii) Since $\mathrm{End}_R(N) = \mathrm{Mat}_2(R)$ is not a Baer ring (see Theorem 2.9), N is not a Baer module by Theorem 2.27.

The following example also illustrates Theorem 3.29.

EXAMPLE 3.31 (see [**15**, Example 4.16]). Let R be a Dedekind domain. When N is an R-module which is not finitely generated, Theorem 3.29 does not hold also. In fact, let $R = \mathbb{Z}$. By Theorem 3.14, there exists a nonempty set Λ such that $\mathrm{CFM}_\Lambda(\mathbb{Z})$ is not a Baer ring because $\mathbb{Z}$ is not semiprimary. Let $N = \mathbb{Z}^{(\Lambda)}$ as a $\mathbb{Z}$-module. Then we have the following.

(i) N is torsion-free.

(ii) N is a Rickart module by Theorem 2.10 and Corollary 2.46. The fact that N is a Rickart module also follows because $\mathbf{Ric}(N) = N$ from Theorem 3.16.

(iii) N is not a Baer module by Theorem 2.27 because $\mathrm{End}(N) = \mathrm{CFM}_\Lambda(\mathbb{Z})$ is not a Baer ring.

We remark that in Example 3.31, N has no Baer hull from Theorem 3.22 because N is not a Baer module. Further, whenever Λ is an uncountable set, N is not a Baer module by [**33**, Remark, p.32].

There are strong connections between quasi-retractable modules and Baer modules (see Theorem 2.32). Thus for a given module N, we may expect that the smallest quasi-retractable essential extension of N (i.e., quasi-retractable hull) and the smallest Baer essential extension (i.e., Baer hull) of N will coincide if they exist (in a fixed injective hull). But the following example eliminates our expectation.

EXAMPLE 3.32. (i) By Example 3.31, there exists a nonempty set Λ such that $N := \mathbb{Z}^{(\Lambda)}$ is not Baer as a $\mathbb{Z}$-module. Further, as the remark of Example 3.31, N has no Baer hull. But note that $\mathbb{Z}_{\mathbb{Z}}^{(\Lambda)}$ itself is the quasi-retractable hull of $\mathbb{Z}_{\mathbb{Z}}^{(\Lambda)}$ by Theorem 2.30(ii).

(ii) (see [**24**, Example 1(iii)]). Let $V = \mathbb{Z}_2 \oplus \mathbb{Z}[1/3]$ as a $\mathbb{Z}$-module. First, we show that V is quasi-retractable. Let $S = \mathrm{End}_{\mathbb{Z}}(M)$. Then

$$S = \begin{bmatrix} \mathrm{End}_{\mathbb{Z}}(\mathbb{Z}_2) & \mathrm{Hom}_{\mathbb{Z}}(\mathbb{Z}[1/3], \mathbb{Z}_2) \\ \mathrm{Hom}_{\mathbb{Z}}(\mathbb{Z}_2, \mathbb{Z}[1/3]) & \mathrm{End}_{\mathbb{Z}}(\mathbb{Z}[1/3]) \end{bmatrix} \cong \begin{bmatrix} \mathbb{Z}_2 & \mathrm{Hom}_{\mathbb{Z}}(\mathbb{Z}[1/3], \mathbb{Z}_2) \\ 0 & \mathbb{Z}[1/3] \end{bmatrix}$$

because $\mathrm{Hom}_{\mathbb{Z}}(\mathbb{Z}_2, \mathbb{Z}[1/3]) = 0$.

Let $a \in \mathbb{Z}$, we use $\overline{a}$ to denote the image of a in $\mathbb{Z}_2$. For description of $\mathrm{Hom}_{\mathbb{Z}}(\mathbb{Z}[1/3], \mathbb{Z}_2)$, we take $f \in \mathrm{Hom}_{\mathbb{Z}}(\mathbb{Z}[1/3], \mathbb{Z}_2)$. Let $f(1) = \alpha$. Then

$$3^m f(1/3^m) = f(1) = \alpha$$

for any non-negative integer m, hence $f(1/3^m) = \alpha$ because $\overline{3^m} = \overline{1}$. Note that

$$\mathbb{Z}[1/3] = \{n/3^m \mid n \in \mathbb{Z} \text{ and } m \text{ is a non-negative integer}\}.$$

Now for $n/3^m \in \mathbb{Z}[1/3]$, $f(n/3^m) = nf(1/3^m) = n\alpha \in \mathbb{Z}_2$. Therefore the image of f is completely determined by $f(1)$. We define

$$\theta : \mathrm{Hom}_{\mathbb{Z}}(\mathbb{Z}[1/3], \mathbb{Z}_2) \to \mathbb{Z}_2 \text{ by } \theta(f) = f(1) \text{ for } f \in \mathrm{Hom}_{\mathbb{Z}}(\mathbb{Z}[1/3], \mathbb{Z}_2).$$

Then θ is an additive group isomorphism. Therefore we may identify S with

$$\begin{bmatrix} \mathbb{Z}_2 & \mathbb{Z}_2 \\ 0 & \mathbb{Z}[1/3] \end{bmatrix} \text{ so that } S = \begin{bmatrix} \mathbb{Z}_2 & \mathbb{Z}_2 \\ 0 & \mathbb{Z}[1/3] \end{bmatrix}.$$

To show that V is quasi-retractable, let I be a left ideal of S with $r_V(I) \neq 0$. Then there exists $0 \neq \beta := \begin{bmatrix} \overline{m_1} \\ m_2 \end{bmatrix} \in V$ with $\overline{m_1} \in \mathbb{Z}_2$ and $m_2 \in \mathbb{Z}[1/3]$ such that $I\beta = 0$. Thus for any $\begin{bmatrix} \overline{x_{11}} & \overline{x_{12}} \\ 0 & x_{22} \end{bmatrix} \in I$, we have that

$$\begin{bmatrix} \overline{x_{11}} & \overline{x_{12}} \\ 0 & x_{22} \end{bmatrix} \begin{bmatrix} \overline{m_1} \\ m_2 \end{bmatrix} = \begin{bmatrix} \overline{x_{11}}\overline{m_1} + \overline{x_{12}}\overline{m_2} \\ x_{22}m_2 \end{bmatrix} = 0.$$

We take

$$s = \begin{bmatrix} 0 & \overline{m_1} \\ 0 & m_2 \end{bmatrix} \in S.$$

Then $s \neq 0$ since $\beta \neq 0$. Now for any $\begin{bmatrix} \overline{x_{11}} & \overline{x_{12}} \\ 0 & x_{22} \end{bmatrix} \in I$,

$$\begin{bmatrix} \overline{x_{11}} & \overline{x_{12}} \\ 0 & x_{22} \end{bmatrix} = \begin{bmatrix} \overline{x_{11}} & \overline{x_{12}} \\ 0 & x_{22} \end{bmatrix} \begin{bmatrix} 0 & \overline{m_1} \\ 0 & m_2 \end{bmatrix} = \begin{bmatrix} 0 & \overline{x_{11}}\overline{m_1} + \overline{x_{12}}\overline{m_2} \\ 0 & x_{22}m_2 \end{bmatrix} = 0$$

by the preceding argument. So $0 \neq s \in r_S(I)$, and hence $r_S(I) \neq 0$. Therefore V is quasi-retractable from the remark after Definition 2.29.

Now let $X = \mathbb{Z}_2 \oplus 2\mathbb{Z} \subseteq V$. Then

$$\ell_S(X) = \begin{bmatrix} 0 & \mathbb{Z}_2 \\ 0 & 0 \end{bmatrix}.$$

Hence there is no $e^2 = e \in S$ such that $\ell_S(X) = Se$. Therefore V is not a Baer $\mathbb{Z}$-module.

However, the Baer hull of V is $\mathbb{Z}_2 \oplus \mathbb{Z}[1/3][1/2] = \mathbb{Z}_2 \oplus \mathbb{Z}[1/6]$ from Theorem 3.22 because $\mathbb{Z}[1/6]$ is Baer as a $\mathbb{Z}$-module. Therefore the quasi-retractable hull of V and the Baer hull of V do not coincide.

(iii) ([**24**, Example 8(ii)]). Let $W = \mathbb{Z}[1/6] \oplus \mathbb{Z}[1/2]$ as a $\mathbb{Z}$-module. Then by Example 3.23,

$$\mathrm{End}_{\mathbb{Z}}(W) \cong \begin{bmatrix} \mathbb{Z}[1/6] & \mathbb{Z}[1/6] \\ 0 & \mathbb{Z}[1/2] \end{bmatrix}.$$

By the method used in part (ii), W is quasi-retractable. Note that W is not a Baer module by Example 3.23. Because $T(\mathbb{Z}) = \sum_{\ell \geq 0} \mathbb{Z}^{-\ell} = \mathbb{Z}$, W has no Baer hull from Theorem 3.22.

The following theorem describes the extending hull of certain modules over a commutative domain.

THEOREM 3.33 ([**23**, Theorem 2.38]). *Assume that R is a commutative domain with the field of fractions F. Let A be an intermediate domain between R and F. Then the following are equivalent.*

(i) A is a Prüfer domain.

(ii) $E(M_R) \oplus A_R^{(n)}$ is the extending hull of $M_R \oplus A_R^{(n)}$ for any R-module M with $Ann_R(M) \neq 0$ and for any positive integer n.

(iii) $A_R^{(2)}$ *is an extending module.*

(iv) $A_R^{(2)}$ *is a Baer module.*

In spite of Theorem 2.39, the following example shows the disparity between Baer hulls and extending hulls for a certain class of modules.

EXAMPLE 3.34 (cf. [**23**, Example 2.42]). Let p be a prime integer in $\mathbb{Z}$. Then we have the following.

(i) $\mathbb{Z}_{p^\infty} \oplus \mathbb{Z}$ is the extending hull of $\mathbb{Z}_p \oplus \mathbb{Z}$ by Theorem 3.33.

(ii) $\mathbb{Z}_p \oplus \mathbb{Z}[1/p]$ is the Baer and the Rickart hull of $\mathbb{Z}_p \oplus \mathbb{Z}$ from Theorem 3.8 and Theorem 3.16.

(iii) $\mathbb{Z}_{p^2} \oplus \mathbb{Z}$ has an extending hull, which is $\mathbb{Z}_{p^\infty} \oplus \mathbb{Z}$ by Theorem 3.33.

(iv) From Theorem 3.8 and Theorem 3.16, the $\mathbb{Z}$-module $\mathbb{Z}_{p^2} \oplus \mathbb{Z}$ has no Baer and no Rickart hull because $\mathbb{Z}_{p^2}$ is not semisimple.

Acknowledgments. The authors appreciate a thorough report from the referee, and are thankful to him/her for helpful comments and suggestions for improvement in the presentation of this work. We are grateful for the partial grant support received from OSU-Lima, the Office of International Affairs, Mathematics Research Institute and the ASC College, The Ohio State University. The authors also thank each other's institutions for the hospitality received during the research work on this paper.

References

[1] R. Baer, *Linear algebra and projective geometry*, Academic Press Inc., New York, NY, 1952. MR0052795

[2] G. F. Birkenmeier, J. K. Park, and S. T. Rizvi, *Extensions of rings and modules*, Birkhäuser/Springer, New York, 2013. MR3099829

[3] B. Blackadar, *Operator algebras: Theory of C^*-algebras and von Neumann algebras*, Encyclopaedia of Mathematical Sciences, vol. 122, Operator Algebras and Non-commutative Geometry, III, Springer-Verlag, Berlin, 2006. MR2188261

[4] S. U. Chase, *A generalization of the ring of triangular matrices*, Nagoya Math. J. **18** (1961), 13–25. MR0123594

[5] A. W. Chatters and C. R. Hajarnavis, *Rings with chain conditions*, Research Notes in Mathematics, vol. 44, Pitman (Advanced Publishing Program), Boston, Mass.-London, 1980. MR590045

[6] A. W. Chatters and S. M. Khuri, *Endomorphism rings of modules over nonsingular CS rings*, J. London Math. Soc. (2) **21** (1980), no. 3, 434–444, DOI 10.1112/jlms/s2-21.3.434. MR577719

[7] M. Fontana, J. A. Huckaba, I. J. Papick, and M. Roitman, *Prüfer domains and endomorphism rings of their ideals*, J. Algebra **157** (1993), no. 2, 489–516, DOI 10.1006/jabr.1993.1112. MR1220780

[8] R. Gilmer, *Multiplicative ideal theory*, Pure and Applied Mathematics, vol. 12, Marcel Dekker, Inc., New York, 1972. MR0427289

[9] A. Hattori, *A foundation of torsion theory for modules over general rings*, Nagoya Math. J. **17** (1960), 147–158. MR0137745

[10] I. Kaplansky, *Projections in Banach algebras*, Ann. of Math. (2) **53** (1951), 235–249, DOI 10.2307/1969540. MR0042067

[11] I. Kaplansky, *Rings of operators*, W. A. Benjamin, Inc., New York-Amsterdam, 1968. MR0244778

[12] I. Kaplansky, *Commutative rings*, Revised edition, The University of Chicago Press, Chicago, Ill.-London, 1974. MR0345945

[13] L.Ya. Kulikov, *On the theory of Abelian groups of arbitrary power*, Mat. Sbornik **9** (1941), 165–182.

[14] T. Y. Lam, *Lectures on modules and rings*, Graduate Texts in Mathematics, vol. 189, Springer-Verlag, New York, 1999. MR1653294

[15] G. Lee, J. K. Park, S. T. Rizvi, and C. S. Roman, *Quasi-Baer module hulls and applications*, J. Pure Appl. Algebra **222** (2018), no. 9, 2427–2455, DOI 10.1016/j.jpaa.2017.09.021. MR3782997

[16] G. Lee, S. T. Rizvi, and C. S. Roman, *Rickart modules*, Comm. Algebra **38** (2010), no. 11, 4005–4027, DOI 10.1080/00927872.2010.507232. MR2764847

[17] G. Lee, S. T. Rizvi, and C. S. Roman, *Direct sums of Rickart modules*, J. Algebra **353** (2012), 62–78, DOI 10.1016/j.jalgebra.2011.12.003. MR2872436

[18] T.-K. Lee and Y. Zhou, *Reduced modules*, Rings, modules, algebras, and abelian groups, Lecture Notes in Pure and Appl. Math., vol. 236, Dekker, New York, 2004, pp. 365–377. MR2050725

[19] H. Lenzing, *Halberbliche Endomorphismenringe* (German), Math. Z. **118** (1970), 219–240, DOI 10.1007/BF01113346. MR0280520

[20] S. Maeda, *On a ring whose principal right ideals generated by idempotents form a lattice*, J. Sci. Hiroshima Univ. Ser. A **24** (1960), 509–525. MR0133348

[21] J. K. Park and S. T. Rizvi, *Baer module hulls of certain modules over a Dedekind domain*, J. Algebra Appl. **15** (2016), no. 8, 1650141, 24 pp., DOI 10.1142/S0219498816501413. MR3528763

[22] J. K. Park and S. T. Rizvi, *On some classes of module hulls*, Algebra and Its Applications, Springer Proc. Math. Stat., vol. 174, Springer, Singapore, 2016, pp. 1–30. MR3613779

[23] J. K. Park and S. T. Rizvi, *Module hulls—similarities and contrasts*, Advances in Rings and Modules, Contemp. Math., vol. 715, Amer. Math. Soc., Providence, RI, 2018, pp. 255–271, DOI 10.1090/conm/715/14416. MR3852393

[24] J. K. Park and S. T. Rizvi, *Properties inherited by direct sums of copies of a module*, Algebra and Its Applications, De Gruyter Proc. Math., De Gruyter, Berlin, 2018, pp. 163–180. MR3837687

[25] D. S. Passman, *A course in ring theory*, The Wadsworth & Brooks/Cole Mathematics Series, Wadsworth & Brooks/Cole Advanced Books & Software, Pacific Grove, CA, 1991. MR1096302

[26] S. T. Rizvi and C. S. Roman, *Baer and quasi-Baer modules*, Comm. Algebra **32** (2004), no. 1, 103–123, DOI 10.1081/AGB-120027854. MR2036224

[27] S. T. Rizvi and C. S. Roman, *On $\mathcal{K}$-nonsingular modules and applications*, Comm. Algebra **35** (2007), no. 9, 2960–2982, DOI 10.1080/00927870701404374. MR2356310

[28] S. T. Rizvi and C. S. Roman, *On direct sums of Baer modules*, J. Algebra **321** (2009), no. 2, 682–696, DOI 10.1016/j.jalgebra.2008.10.002. MR2483287

[29] J. J. Rotman, *An introduction to homological algebra*, Pure and Applied Mathematics, vol. 85, Academic Press, Inc. [Harcourt Brace Jovanovich, Publishers], New York-London, 1979. MR538169

[30] W. R. Scott, *Group theory*, 2nd ed., Dover Publications, Inc., New York, 1987. MR896269

[31] D. W. Sharpe and P. Vámos, *Injective modules*, Cambridge Tracts in Mathematics and Mathematical Physics, vol. 62, Cambridge University Press, London-New York, 1972. MR0360706

[32] W. Stephenson and G.M. Tsukerman, *Rings of endomorphisms of projective modules*, Siberian Math. J. **11** (1970), 181–184 (English translation).

[33] G. V. Wilson, *Modules with the summand intersection property*, Comm. Algebra **14** (1986), no. 1, 21–38, DOI 10.1080/00927878608823297. MR814137

Department of Mathematics, Busan National University, Busan 46241, South Korea
Email address: jkp1128@yahoo.com

Department of Mathematics, The Ohio State University, Lima, OH 45804-3576, U.S.A.
Email address: rizvi.1@osu.edu

Contemporary Mathematics
Volume **738**, 2019
https://doi.org/10.1090/conm/738/14881

Tableau correspondences and representation theory

Digjoy Paul, Amritanshu Prasad, and Arghya Sadhukhan

ABSTRACT. We deduce decompositions of natural representations of general linear groups and symmetric groups from combinatorial bijections involving tableaux. These include some of Howe's dualities, Gelfand models, the Schur-Weyl decomposition of tensor space, and multiplicity-free decompositions indexed by threshold partitions.

1. Introduction

Bijections in algebraic combinatorics are often manifestations of results in representation theory. For example, the Robinson-Schensted correspondence, between permutations and pairs of standard tableaux of the same shape, reflects the fact that the sum of squares of dimensions of irreducible representations of a symmetric group is its order. Sometimes combinatorial identities can be used to *prove* results in representation theory. For instance, the classification of the irreducible representations of symmetric groups and Young's rule are deduced from the Robinson-Schensted-Knuth (RSK) correspondence in [**12**]. The dual RSK correspondence is used to determine what happens when a representation is twisted by the sign character. This article collects many more instances of this phenomenon.

Section 2 reviews basic facts about semistandard Young tableaux and Schur functions. Section 3 reviews fundamental results in the polynomial representation theory of general linear groups. The RSK correspondence is a bijection from matrices with non-negative integer entries onto pairs of semistandard tableaux of the same shape. The dual RSK correspondence is a bijection from matrices with entries 0 or 1 onto pairs of semistandard tableaux of mutually conjugate shape. In Section 4, the RSK correspondence and its dual are shown to imply the $(\mathrm{GL}_m, \mathrm{GL}_n)$-duality and the skew $(\mathrm{GL}_m, \mathrm{GL}_n)$-duality theorems of Howe. The symmetry property of the RSK correspondence gives a Gelfand model for $\mathrm{GL}_n(\mathbf{C})$. Schützenberger's lemma on the RSK correspondence is shown to give a refinement of the Gelfand model.

2010 *Mathematics Subject Classification.* 05E10,20C30,22E46.

Key words and phrases. Burge correspondence, Gelfand model, Schur-Weyl duality, Specht modules, RSK correspondence, tableaux, Weyl modules.

The second author is the corresponding author.

AP was supported by Swarnajayanti Fellowship DST/SJF/MSA-02/2014-15, Department of Science & Technology, Government of India.

The Burge correspondence is another bijection from matrices with non-negative integer entries onto pairs of semistandard tableaux of the same shape, less well-known than the RSK correspondence. In Section 5, we show that its symmetry property gives rise to another Gelfand model for $\mathrm{GL}_n(\mathbf{C})$, which also has a refinement based on an adaptation of Schützenberger's lemma to this setting. Two more correspondences of Burge give rise to multiplicity-free decompositions into representations parameterized by threshold partitions, and conjugate threshold partitions. These are discussed in Section 6.

Section 7 describes a passage from representations of $\mathrm{GL}_n(\mathbf{C})$ to representations of S_n via the all-ones weight space. Applying this to one side of $(\mathrm{GL}_m, \mathrm{GL}_n)$-duality, we recover Schur-Weyl duality. Applying it to the Gelfand models of $\mathrm{GL}_n(\mathbf{C})$, we recover combinatorial Gelfand models for S_n due to Inglis, Richardson, and Saxl [**8**]. Applying it to the multiplicity-free representations of $\mathrm{GL}_n(\mathbf{C})$ in Section 6, we recover some other interesting multiplicity-free representations of S_n, which were outlined by Bump [**3**] using completely different methods, and are given by Macdonald [**11**, Ch. I.5, Example 9].

Most of the results in this article have appeared in the MSc thesis of Arghya Sadhukhan [**14**].

2. Semistandard Young Tableaux and Schur Polynomials

In this section we recall basic facts about semistandard Young tableaux and Schur polynomials. For detailed, self-contained expositions, see [**11**–**13**].

Definition 2.1. A *semistandard Young tableau* in n letters is a left-justified array of boxes with rows of weakly decreasing length filled with numbers between 1 and n such that

(1) the numbers increase weakly from left to right along rows,
(2) the numbers increase strictly from top to bottom along columns.

For brevity, a semistandard Young tableau will be referred to simply as a *tableau* in this article. If a tableau has l rows, and λ_i boxes in the ith row for $i = 1, \dots, l$, then the partition $\lambda = (\lambda_1, \dots, \lambda_l)$ is called the shape of the tableau. The weight of a tableau is the integer vector $(\mu_1, \dots, \mu_n)$ where μ_i is the number of times the number i occurs. If t is a tableau, we write $\mathrm{wt}(t)$ for its weight. Write $\mathrm{Tab}_n(\lambda)$ for the set of all tableaux in n letters having shape λ. The subset of $\mathrm{Tab}_n(\lambda)$ consisting of tableaux with weight μ is denoted by $\mathrm{Tab}(\lambda, \mu)$.

Example 2.2. The tableau in 7 letters

$$t = \begin{array}{|c|c|c|c|c|} \hline 2 & 2 & 3 & 5 & 5 \\ \hline 4 & 4 & 4 & 6 \\ \cline{1-4} 5 & 7 \\ \cline{1-2} \end{array}$$

has shape $(5, 4, 2)$ and $\mathrm{wt}(t) = (0, 2, 1, 3, 3, 1, 1)$.

Given an integer vector $\mu = (\mu_1, \dots, \mu_n)$ with n-coordinates, let x^μ denote the monomial $x_1^{\mu_1} \cdots x_n^{\mu_n}$ in n variables.

Definition 2.3 (Schur Polynomial). For each integer partition λ, the Schur polynomial in n variables corresponding to λ is defined as:

$$s_\lambda(x_1, \dots, x_n) = \sum_{t \in \mathrm{Tab}_n(\lambda)} x^{\mathrm{wt}(t)}.$$

It turns out that s_λ is a symmetric polynomial in n variables. If the partition λ has more than n non-zero parts, there are no tableaux in n letters of shape λ (because the first column of the tableau has to be strictly increasing, and has length equal the number of parts of λ). On the other hand, if the number of non-zero parts of λ is at most n, then there is at least one such tableau, namely the one whose ith row has all boxes filled with i. Therefore $s_\lambda(x_1, \dots, x_n) = 0$ if and only if λ has more than n non-zero parts.

Let Λ_n^d denote the set of all integer partitions of d with at most n parts. By padding a partition with 0's, write it as an integer vector $\lambda = (\lambda_1, \dots, \lambda_n)$ with exactly n parts. Thus

$$\Lambda_n^d = \{(\lambda_1, \dots, \lambda_n) \in \mathbf{Z}^n \mid \lambda_1 + \cdots + \lambda_n = d,\ \lambda_1 \geq \cdots \geq \lambda_n \geq 0\}.$$

THEOREM 2.4 ([**11**, Ch. I,(3.2)], [**12**, Theorem 5.4.3]). *The set*

$$\{s_\lambda \mid \lambda \in \Lambda_n^d\}$$

is a basis for the space of all symmetric polynomials of degree d in n variables with coefficients in $\mathbf{C}$.

3. Polynomial Representations of $\mathrm{GL}_n(\mathbf{C})$

In this section we recall basic facts about polynomial representations of $\mathrm{GL}_n(\mathbf{C})$. Details can be found in [**12**, Chapter 6].

DEFINITION 3.1. A *polynomial representation* of $\mathrm{GL}_n(\mathbf{C})$ is a finite dimensional vector space W over $\mathbf{C}$, together with a group homomorphism $\rho : \mathrm{GL}_n(\mathbf{C}) \to \mathrm{GL}(W)$, such that each entry of the matrix of $\rho(g)$ (with respect to any basis of W) is a polynomial in the entries of the matrix g. The polynomial representation W is said to be *homogeneous of degree d* if these polynomials are homogeneous of degree d. It is said to be *irreducible* if there is no non-trivial proper subspace $W' \subset W$ that is left invariant by $\rho(g)$ for all $g \in \mathrm{GL}_n(\mathbf{C})$. Representations (ρ_1, W_1) and (ρ_2, W_2) are said to be *isomorphic* if there exists a linear isomorphism $T : W_1 \to W_2$ such that $\rho_2(g) \circ T = T \circ \rho_1(g)$ for each $g \in \mathrm{GL}_n(\mathbf{C})$.

EXAMPLE 3.2 (The trivial representation). The only homogeneous polynomial representations of degree 0 are the ones where $\rho(g)$ is the identity operator of W for every $g \in \mathrm{GL}_n(\mathbf{C})$.

EXAMPLE 3.3 (Direct sum). If (ρ_1, W_1) and (ρ_2, W_2) are polynomial representations of $\mathrm{GL}_n(\mathbf{C})$, then $(\rho_1 \oplus \rho_2, W_1 \oplus W_2)$ is also a polynomial representation of $\mathrm{GL}_n(\mathbf{C})$. If ρ_1 and ρ_2 are homogeneous of degree d, then so is $\rho_1 \oplus \rho_2$.

EXAMPLE 3.4 (The defining representation). The identity map $\rho : \mathrm{GL}_n(\mathbf{C}) \to \mathrm{GL}_n(\mathbf{C})$ defines a representation of $\mathrm{GL}_n(\mathbf{C})$ on $\mathbf{C}^n$. This representation is homogeneous of degree 1 and is called the *defining representation.*

EXAMPLE 3.5 (Tensor product). If (ρ_1, W_1) and (ρ_2, W_2) are polynomial representations of $\mathrm{GL}_n(\mathbf{C})$, then so is $(\rho_1 \otimes \rho_2, W_1 \otimes W_2)$. If ρ_1 has degree d_1 and ρ_2 has degree d_2, then $\rho_1 \otimes \rho_2$ has degree $d_1 + d_2$. As a particular example, $\otimes^d \mathbf{C}^n$, the d-fold tensor power of the defining representation $\mathbf{C}^n$, is homogeneous of degree d and dimension n^d.

EXAMPLE 3.6 (Symmetric and alternating tensors). The symmetric group S_d acts on $\otimes^d \mathbf{C}^n$ by permuting the tensor factors. The subspace of invariant tensors is denoted by $\mathrm{Sym}^d \mathbf{C}^n$. Since the actions of $\mathrm{GL}_n(\mathbf{C})$ and S_d on $\otimes^d \mathbf{C}^n$ commute, $\mathrm{Sym}^d \mathbf{C}^n$ is invariant under the action of $\mathrm{GL}_n(\mathbf{C})$, forming a homogeneous representation of degree d, which is known as the dth symmetric tensor. Let L denote the subalgebra of $\bigoplus_{d=0}^{\infty} \otimes^d \mathbf{C}^n$ generated by $x \otimes x$. Let $L_d = L \cap \otimes^d \mathbf{C}^n$. Then L_d is an invariant subspace of $\otimes^d \mathbf{C}^n$, and the quotient $\wedge^d \mathbf{C}^n = \otimes^d \mathbf{C}^n / L_d$ is a representation of $\mathrm{GL}_n(\mathbf{C})$, known as the dth alternating tensor.

Let $\lambda = (\lambda_1, \dots, \lambda_l)$ be a partition of d. Then

$$\mathrm{Sym}^\lambda \mathbf{C}^n = \bigotimes_{i=1}^{l} \mathrm{Sym}^{\lambda_i} \mathbf{C}^n$$

$$\wedge^\lambda \mathbf{C}^n = \bigotimes_{i=1}^{l} \wedge^{\lambda_i} \mathbf{C}^n$$

are homogeneous polynomial representations of $\mathrm{GL}_n(\mathbf{C})$ of degree d.

Let $T_n \subset \mathrm{GL}_n(\mathbf{C})$ denote the subgroup of invertible diagonal matrices. We use $\Delta(x_1, \dots, x_n)$ to denote the diagonal matrix with entries $x_1, \dots, x_n$.

DEFINITION 3.7 (Weight vector). Let (ρ, W) be a polynomial representation of $\mathrm{GL}_n(\mathbf{C})$. A vector $v \in W$ is said to be a *weight vector with weight* $\mu = (\mu_1, \dots, \mu_n)$ if, for all $x_1, \dots, x_n \in \mathbf{C}^*$,

$$\rho(\Delta(x_1, \dots, x_n))v = x_1^{\mu_1} \cdots x_m^{\mu_n} v.$$

The subspace of all weight vectors of weight μ is called the *μ-weight space* of W and denoted $W(\mu)$.

THEOREM 3.8 ([**11**, Appendix A, (2.4)], [**12**, Theorem 6.6.5]). *Every polynomial representation (ρ, W) of $\mathrm{GL}_n(\mathbf{C})$ admits a basis of weight vectors.*

EXAMPLE 3.9. In the defining representation of $\mathrm{GL}_n(\mathbf{C})$ (Example 3.4), the coordinate vector $e_i \in \mathbf{C}^n$ is a weight vector with weight e_i. The set $\{e_i \mid i = 1, \dots, n\}$ is a basis of weight vectors for $\mathbf{C}^n$.

EXAMPLE 3.10. In the tensor space $\otimes^d \mathbf{C}^n$ (Example 3.5),

$$\{e_{i_1} \otimes \cdots \otimes e_{i_d} \mid (i_1, \dots, i_d) \in \{1, \dots, n\}^d\}$$

is a basis of $\otimes^d \mathbf{C}^n$ consisting of weight vectors. The weight of $e_{i_1} \otimes \cdots \otimes e_{i_d}$ is $(\mu_1, \dots, \mu_n)$, where μ_k is the number of k for which $i_k = i$.

EXAMPLE 3.11. For non-negative integers n and d, define

$$M(n, d) = \{(i_1, \dots, i_d) \mid 1 \leq i_1 \leq \cdots \leq i_d \leq n\}.$$

Given $I = (i_1, \dots, i_d) \in M(n, d)$ the symmetric tensor $e_{i_1} \cdots e_{i_d}$ is a weight vector in $\mathrm{Sym}^d \mathbf{C}^n$ with weight $(\mu_1, \dots, \mu_n)$ where μ_i is the number of k for which $i_k = i$. As I runs over $M(n, d)$, these vectors form a basis of $\mathrm{Sym}^d \mathbf{C}^n$.

For non-negative integers n and d, define

$$N(n, d) = \{(i_1, \dots, i_d) \mid 1 \leq i_1 < \cdots < i_d \leq n\}.$$

Given $I = (i_1, \dots, i_d) \in N(n, d)$ the alternating tensor $e_{i_1} \wedge \cdots \wedge e_{i_d}$ is a weight vector in $\wedge^d \mathbf{C}^n$ with weight $(\mu_1, \dots, \mu_n)$ where μ_i is the number of k for which $i_k = i$. As I runs over $N(n, d)$, these vectors form a basis of $\wedge^d \mathbf{C}^n$.

Most proofs in this article will be based on finding bases of weight vectors of representations as in the examples above.

DEFINITION 3.12. The *character* of a polynomial representation W of $\mathrm{GL}_n(\mathbf{C})$ is defined as

$$\mathrm{ch}_W(x_1, \ldots, x_n) = \mathrm{tr}(\rho(\Delta(x_1, \ldots, x_n)); W).$$

For any permutation $w \in S_n$, $\Delta(x_{w(1)}, \ldots, x_{w(n)})$ is conjugate to $\Delta(x_1, \ldots, x_n)$. Therefore ch_W is a symmetric polynomial. If W is homogeneous of degree d, then ch_W is homogeneous of degree d.

REMARK 3.13. Theorem 3.8 implies that

$$\mathrm{ch}_W(x_1, \ldots, x_n) = \sum_{\mu} \dim W(\mu) x^{\mu},$$

the sum being over all n-tuples μ of non-negative integers.

Let λ' denote the partition *conjugate* to λ, namely,

$$\lambda'_i = \#\{j \mid \lambda_j \geq i\} \text{ for } i = 1, \ldots, n.$$

THEOREM 3.14 ([**11**, Appendix A, (8.2), (8.4)], [**12**, Chapter 6]). *Let $\rho : \mathrm{GL}_n(\mathbf{C}) \to \mathrm{GL}(W)$ be a polynomial representation of $\mathrm{GL}_n(\mathbf{C})$. Then we have:*

(1) *For every partition $\lambda \in \Lambda^d_n$, there exists a unique irreducible homogeneous polynomial representation W^n_λ of $\mathrm{GL}_n(\mathbf{C})$ of degree d which occurs in both $\mathrm{Sym}^\lambda \mathbf{C}^n$ and $\wedge^{\lambda'} \mathbf{C}^n$. This representation satisfies:*

$$\mathrm{ch}_{W^n_\lambda}(x_1, \ldots, x_n) = s_\lambda(x_1, \ldots, x_n),$$

the Schur polynomial corresponding to λ, in n variables.

(2) *Every polynomial representation W of $\mathrm{GL}_n(\mathbf{C})$ has a unique decomposition of the form:*

$$W = \bigoplus_{\lambda} (W^n_\lambda)^{\oplus m_\lambda}$$

into irreducible polynomial representations (since W is finite dimensional, it should be understood that m_λ is positive for only finitely many λ).

(3) *Two polynomial representations of $\mathrm{GL}_n(\mathbf{C})$ are isomorphic if and only if their characters are equal.*

4. The RSK Correspondence and its Dual

Let $\mu = (\mu_1, \ldots, \mu_m)$ and $\nu = (\nu_1, \ldots, \nu_n)$ be non-negative integer vectors with common sum d. Let $\mathbf{M}_{\mu\nu}$ denote the set of all $m \times n$ matrices $A = (a_{ij})$ with non-negative integer entries such that

$$\sum_j a_{ij} = \mu_i \text{ for } i = 1, \ldots, m \text{ and } \sum_i a_{ij} = \nu_j \text{ for } j = 1, \ldots, n.$$

The RSK correspondence [**9**, Section 3] is an algorithmic bijection:

$$\mathrm{RSK} : \mathbf{M}_{\mu\nu} \tilde{\to} \coprod_{\lambda \vdash d} \mathrm{Tab}(\lambda, \nu) \times \mathrm{Tab}(\lambda, \mu).$$

THEOREM 4.1. *Let $W = \mathrm{Sym}^d(\mathbf{C}^m \otimes \mathbf{C}^n)$. Viewing $\mathbf{C}^m$ and $\mathbf{C}^n$ as the defining representations of $\mathrm{GL}_m(\mathbf{C})$ and $\mathrm{GL}_n(\mathbf{C})$ respectively, the functorial nature of Sym^d implies that W is a representation of $\mathrm{GL}_m(\mathbf{C}) \times \mathrm{GL}_n(\mathbf{C})$. The decomposition of W into irreducible representations of $\mathrm{GL}_m(\mathbf{C}) \times \mathrm{GL}_n(\mathbf{C})$ is given by:*

$$W = \bigoplus_{\lambda \vdash d} W_\lambda^n \otimes W_\lambda^m.$$

PROOF. Let $e_1 \dots, e_n$ and $f_1, \dots, f_m$ denote the coordinate vectors in $\mathbf{C}^n$ and $\mathbf{C}^m$ respectively. Let μ and ν be non-negative integer vectors with common sum d. Given $A \in \mathbf{M}_{\mu\nu}$, define a vector $v_A \in W$ by

$$v_A = \prod_{i=1}^n \prod_{j=1}^m (e_i \otimes f_j)^{a_{ij}}.$$

Then v_A is an eigenvector for the action of $\Delta(x_1, \dots, x_n) \in T_n$, and $\Delta(y_1, \dots, y_m) \in T_m$ with eigenvalue $x^\nu y^\mu$. Therefore

$$\begin{aligned}
\mathrm{tr}(\Delta(x)\Delta(y); W) &= \sum_\nu \sum_\mu \sum_{A \in \mathbf{M}_{\mu\nu}} x^\nu y^\mu \\
&= \sum_\nu \sum_\mu \sum_\lambda \sum_{t' \in \mathrm{Tab}(\lambda,\nu)} \sum_{t'' \in \mathrm{Tab}(\lambda,\mu)} x^{\mathrm{wt}(t')} y^{\mathrm{wt}(t'')} \\
&= \sum_\lambda s_\lambda(x) s_\lambda(y) \\
&= \sum_\lambda \mathrm{ch}_{W_\lambda^n}(x) \mathrm{ch}_{W_\lambda^m}(y) \\
&= \sum_\lambda \mathrm{ch}_{W_\lambda^n \otimes W_\lambda^m}(x, y).
\end{aligned}$$

□

REMARK 4.2. The equality of characters in above Theorem is the equality of the homogeneous parts of degree d in the Cauchy identity (see [**11**, Ch. I, (4.3)])

$$\prod_{i,j \geq 1} (1 - x_i y_j)^{-1} = \sum_\lambda s_\lambda(x) s_\lambda(y).$$

Theorem 4.1 is called $(\mathrm{GL}_n, \mathrm{GL}_m)$-*duality* by Howe [**7**, Theorem 2.1.2], who gives a different proof. See also Benson and Ratcliff [**1**, Theorem 4.1.1].

Let $\mu = (\mu_1, \dots, \mu_m)$ and $\nu = (\nu_1, \dots, \nu_n)$ be non-negative integer vectors with common sum d. Let $\mathbf{N}_{\mu\nu}$ denote the set of all $m \times n$ matrices $A = (a_{ij})$ with entries in $\{0, 1\}$ such that

$$\sum_j a_{ij} = \mu_i \text{ for } i = 1, \dots, m \text{ and } \sum_i a_{ij} = \nu_j \text{ for } j = 1, \dots, n.$$

The dual RSK correspondence [**9**, Section 5] is an algorithmic bijection:

$$\mathrm{RSK}^* : \mathbf{N}_{\mu\nu} \tilde{\to} \coprod_{\lambda \vdash d} \mathrm{Tab}(\lambda', \nu) \times \mathrm{Tab}(\lambda, \mu).$$

THEOREM 4.3. *Let* $W = \wedge^d(\mathbf{C}^m \otimes \mathbf{C}^n)$. *The decomposition of* W *into irreducible representations of* $\mathrm{GL}_m(\mathbf{C}) \times \mathrm{GL}_n(\mathbf{C})$ *is given by:*

$$W = \bigoplus_{\lambda} W_{\lambda'}^n \otimes W_\lambda^m.$$

PROOF. Given A in $\mathbf{N}_{\mu\nu}$ define $v_A \in W$ by:

$$v_A = \bigwedge_{i=1}^{n} \bigwedge_{j=1}^{m} (e_i \otimes f_j)^{a_{ij}}.$$

Then v_A is an eigenvector for the action of $\Delta(x_1, \ldots, x_n) \in T_n$ and $\Delta(y_1, \ldots, y_m) \in T_m$ with eigenvalue $x^\nu y^\mu$. Now, proceeding as in the proof of Theorem 4.1 gives the result. □

REMARK 4.4. The corresponding equality of characters is the Cauchy identity (see [**11**, Ch. I, (4.3′)])

$$\prod_{i,j \geq 1} (1 + x_i y_j) = \sum_{\lambda} s_\lambda(x) s_{\lambda'}(y).$$

Theorem 4.3 is called skew $(\mathrm{GL}_n, \mathrm{GL}_m)$-*duality* by Howe [**7**, Theorem 4.1.1], who gives a different proof.

Knuth proved a symmetry theorem for the RSK correspondence:

$$\mathrm{RSK}(A) = (P, Q) \text{ if and only if } \mathrm{RSK}(A') = (Q, P).$$

Here A' denotes the transpose of the matrix A. Therefore, if A is symmetric, then $\mathrm{RSK}(A)$ is of the form (P, P) for some semistandard Young tableau P. Let $\mathbf{M}_{\nu\nu}^{\mathrm{sym}}$ denote the set of symmetric $n \times n$ matrices in $\mathbf{M}_{\nu\nu}$. The symmetry property of the RSK correspondence implies that it induces a bijection:

$$\mathrm{RSK} : \mathbf{M}_{\nu\nu}^{\mathrm{sym}} \tilde{\to} \coprod_{\lambda \vdash d} \mathrm{Tab}(\lambda, \nu). \tag{1}$$

THEOREM 4.5. *Let* $W = \bigoplus_{k+2l=d} \mathrm{Sym}^k(\mathbf{C}^n) \otimes \mathrm{Sym}^l(\wedge^2 \mathbf{C}^n)$. *This representation of* $\mathrm{GL}_n(\mathbf{C})$ *has decomposition into irreducibles given by:*

$$W = \bigoplus_{\lambda \vdash d} W_\lambda^n.$$

PROOF. As A runs over symmetric $n \times n$ matrices with non-negative integer entries summing to d, the vectors

$$v_A = \prod_{i=1}^{n} e_i^{a_{ii}} \prod_{i<j} (e_i \wedge e_j)^{a_{ij}}$$

form a basis of W consisting of eigenvectors for the action of $\Delta(x_1, \ldots, x_n)$. When $A \in M_{\nu\nu}^{\mathrm{sym}}$, the eigenvalue corresponding to v_A is x^ν. Now using the correspondence (1) on symmetric matrices:

$$\begin{aligned}\mathrm{ch}_W(x_1,\dots,x_n) &= \sum_\nu |\mathbf{M}^{\mathrm{sym}}_{\nu\nu}| x^\nu \\ &= \sum_\nu \sum_\lambda |\mathrm{Tab}(\lambda,\nu)| x^\nu \\ &= \sum_\lambda s_\lambda(x) \\ &= \sum_\lambda \mathrm{ch}_{W_\lambda}(x),\end{aligned}$$

thereby proving the result. □

REMARK 4.6. The equality of characters above corresponds to the Littlewood identity [**11**, Ch I.5, Example 4]

$$\sum_\lambda s_\lambda = \prod_i (1-x_i)^{-1} \prod_{i<j} (1-x_ix_j)^{-1}.$$

REMARK 4.7 (A Gelfand Model). We may say that the symmetric algebra:

$$\mathrm{Sym}(\mathbf{C}^n \oplus \wedge^2\mathbf{C}^n) := \bigoplus_{k,l\geq 0} \mathrm{Sym}^k(\mathbf{C}^n) \otimes \mathrm{Sym}^l(\wedge^2\mathbf{C}^n)$$

is a *model* for polynomial representation theory of $\mathrm{GL}_n(\mathbf{C})$, in the sense of Bernstein, Gelfand and Gelfand [**2**]: it is a completely reducible representation containing each irreducible polynomial representation of $\mathrm{GL}_n(\mathbf{C})$ exactly once. For an alternate approach, see [**1**, Theorem 4.5.1].

The following result was proved for permutation matrices by Schützenberger [**15**, Theorem 4.4]. An elegant proof using the light-and-shadows version of the Robinson-Schensted correspondence was suggested (as an exercise) by Viennot [**17**, Proposition 4.1]. This proof can be adapted to integer matrices using Fulton's matrix ball construction [**5**, Section 4.2], or the generalization of Viennot's light-and-shadows algorithm in Prasad [**12**, Chapter 3].

THEOREM 4.8 (Schützenberger's lemma). *The RSK correspondence takes symmetric matrices with trace k to semistandard Young tableaux with k odd columns.*

The term "odd columns" alludes to the Young diagram of λ, whose rows are the parts of λ. The columns of λ are the parts of the conjugate partition λ'. Schützenberger's lemma gives the following refinement of Theorem 4.5:

THEOREM 4.9. *For all non-negative integers k and l,*

$$\mathrm{Sym}^k(\mathbf{C}^n) \otimes \mathrm{Sym}^l(\wedge^2\mathbf{C}^n) = \bigoplus_{\lambda\vdash k+2l \text{ with } k \text{ odd columns}} W^n_\lambda.$$

Taking $k=0$ gives a better-known special case (see Benson and Ratcliff [**1**, Theorem 4.3.1])

THEOREM 4.10. *For every positive integer l,*

$$\mathrm{Sym}^l(\wedge^2\mathbf{C}^n) = \bigoplus_{\lambda\vdash 2l \text{with every part appearing even number of times}} W^n_\lambda.$$

REMARK 4.11. The restriction of the RSK correspondence to symmetric matrices with zeros on the diagonal is the correspondence concerning *graphs without loops* in Burge [**4**, Section 2]. Theorem 4.10 can be deduced from it. The identity of characters in Theorem 4.9 comes from equating the coefficients of t^k in the Littlewood identity (see [**11**, Ch. I.5, Example 5(b)], [**16**, Theorem 3.27(2)])

$$\prod_i (1-tx_i)^{-1} \prod_{i<j} (1-x_ix_j)^{-1} = \sum_\lambda s_\lambda(x) t^{\#\text{ columns of odd length in }\lambda}.$$

Setting $t = 0$ corresponds to the character identity of Theorem 4.10. See also [**16**, Theorem 3.30].

5. The Burge Correspondence

Burge [**4**] described four variants of the RSK correspondence. We begin with a correspondence which, although absent from Burge's paper, is now commonly known as *the Burge Correspondence* [**5**, A.4.1]:

$$\mathrm{BUR} : \mathbf{M}_{\mu\nu} \to \coprod_{\lambda \vdash d} \mathrm{Tab}(\lambda, \nu) \times \mathrm{Tab}(\lambda, \mu).$$

Indeed, this is a *different* bijection between the same two sets between which the RSK correspondence defines a bijection. This correspondence shares the symmetry property with the RSK correspondence:

$$\mathrm{BUR}(A) = (P, Q) \text{ if and only if } \mathrm{BUR}(A') = (Q, P),$$

and consequently induces a bijection:

$$\mathrm{BUR} : \mathbf{M}^{\mathrm{sym}}_{\mu\mu} \tilde{\to} \coprod_{\lambda \vdash d} \mathrm{Tab}(\lambda, \mu). \tag{2}$$

An important difference between the Burge correspondence and the RSK correspondence is that Theorem 4.8 (Schützenberger's lemma) fails. Instead we have [**5**, A4.1, Exercise 18]:

THEOREM 5.1 (Schützenberger's lemma for the Burge correspondence). *The Burge correspondence (2) takes a symmetric integer matrix with k odd entries on its diagonal to a semistandard Young tableaux with k odd rows.*

THEOREM 5.2. *For non-negative integers k and l, the representation:*

$$W = \wedge^k(\mathbf{C}^n) \otimes \mathrm{Sym}^l(\mathrm{Sym}^2\mathbf{C}^n)$$

of $\mathrm{GL}_n(\mathbf{C})$ has decomposition into irreducible representations given by:

$$W = \bigoplus_{\lambda \vdash k+2l \text{ with } k \text{ odd rows}} W^n_\lambda.$$

PROOF. Given a symmetric integer matrix A with k odd diagonal entries, write $A = A' + A''$, where A' is a diagonal matrix with diagonal entries 0 or 1, and A'' is an integer matrix with even diagonal entries. Write $b_{ii} = a''_{ii}/2$ (half of the (i,i)th entry of A'') and $b_{ij} = a''_{ij}$ for $i \neq j$. Set

$$v_A = \bigwedge_{a'_{ii}=1} e_i \otimes \prod_{i \leq j} (e_ie_j)^{b_{ij}}.$$

As A runs over symmetric matrices with non-negative integer entries, k odd entries on the diagonal, and off-diagonal entries summing to $2l$, v_A forms a basis of eigenvectors for the action of $\Delta(x_1, \dots, x_n)$ on W. Moreover, when $A \in \mathbf{M}^{\mathrm{sym}}_{\mu\mu}$ then v_A has eigenvalue x^μ.

Schützenberger's lemma for the Burge correspondence (Theorem 5.1) gives:

$$\mathrm{ch}_W(x) = \sum_{\lambda \vdash k+2l \text{ with } k \text{ odd rows}} s_\lambda(x),$$

whence the theorem follows. □

When restricted to matrices with even diagonal entries, the correspondence (2) becomes the correspondence in [**4**, Section 3] concerning *graphs with loops and multiple edges*. In terms of Theorem 5.2, this is the special case where $k = 0$:

THEOREM 5.3. *For every non-negative integer k,*

$$\mathrm{Sym}^k(\mathrm{Sym}^2 \mathbf{C}^n) = \bigoplus_{\lambda \vdash 2k \textit{with all parts even}} W^n_\lambda.$$

REMARK 5.4. The identity of characters in Theorem 5.2 corresponds to the equality of coefficients of t^k in the identity [**11**, Ch I.5, Example 8]:

$$\prod_i \frac{1 + tx_i}{1 - x_i^2} \prod_{i<j} (1 - x_i x_j)^{-1} = \sum_\lambda t^{\#\text{ odd rows in } \lambda} s_\lambda(x).$$

The identity of characters in Theorem 5.3 is obtained obtained by setting $t = 0$ (see [**11**, Ch. I.5, Example 5(a)], [**16**, Theorem 3.34]). Theorem 5.3 appears in Howe [**7**, Theorem 3.1] with a different proof.

REMARK 5.5 (Another Gelfand model). We get a second model (see Remark 4.7) for the polynomial representation theory of $\mathrm{GL}_n(\mathbf{C})$, namely:

$$\bigoplus_{k,l \geq 0} \wedge^k \mathbf{C}^n \otimes \mathrm{Sym}^l(\mathrm{Sym}^2 \mathbf{C}^n).$$

6. Two more correspondences of Burge

Besides the restrictions of RSK and BUR to symmetric matrices with zeros on the diagonal, the article of Burge [**4**] contains two more correspondences. Let $\mathbf{N}^{\mathrm{sym}}_{\mu\mu}$ denote matrices in $\mathbf{M}^{\mathrm{sym}}_{\mu\mu}$ with entries in $\{0, 1\}$. Let $\mathbf{N}^{\mathrm{sym,tr}=0}_{\mu\mu}$ denote the subset of $\mathbf{N}^{\mathrm{sym}}_{\mu\mu}$ consisting of matrices with trace zero. Given a partition λ, for each i such that $\lambda_i \geq i$, let $\alpha_i = \lambda_i - i$, and $\beta_i = \lambda'_i - i$. The largest index f such that $\lambda_f \geq f$ is called the *Frobenius rank* of λ, and $(\alpha_1, \dots, \alpha_f | \beta_1, \dots, \beta_f)$ are called the *Frobenius coordinates* of λ.

DEFINITION 6.1 (Threshold Partition). A partition λ is called a *threshold partition* if its Frobenius coordinates satisfy $\beta_i = \alpha_i + 1$ for $i = 1, \dots, f$, where f is the Frobenius rank of λ.

Let $\mathrm{TP}(n)$ denote the set of threshold partitions of n. Let μ be a partition of d. The correspondences of Burge concerning *graphs without loops or multiple edges*

[**4**, Section 4] and *graphs without multiple edges* [**4**, Section 5] are:

$$\mathrm{BUR}_1 : N_{\mu\mu}^{\mathrm{sym,tr}=0} \tilde{\to} \coprod_{\lambda \in \mathrm{TP}(d)} \mathrm{Tab}(\lambda, \mu)$$

$$\mathrm{BUR}_2 : N_{\mu\mu}^{\mathrm{sym}} \tilde{\to} \coprod_{\lambda' \in \mathrm{TP}(d)} \mathrm{Tab}(\lambda, \mu).$$

Their significance in terms of the representation theory of $\mathrm{GL}_n(\mathbf{C})$ are given by [**11**, Example in Appendix A.7]:

THEOREM 6.2. *For every non-negative integer d, we have:*

$$\wedge^d(\wedge^2 \mathbf{C}^n) \cong \bigoplus_{\lambda \in \mathrm{TP}(d)} W_\lambda^n,$$

$$\wedge^d(\mathrm{Sym}^2 \mathbf{C}^n) \cong \bigoplus_{\lambda' \in \mathrm{TP}(d)} W_\lambda^n.$$

PROOF. Given a symmetric matrix A with entries in $\{0, 1\}$, define

$$v_A = \bigwedge_{\{(i,j) | i<j, \; a_{ij}=1\}} e_i \wedge e_j,$$

taking the terms in lexicographic order. As A runs over symmetric matrices with entries in $\{0, 1\}$, zeros on the diagonal, and all entries summing to d, v_A forms a basis of $\wedge^d(\wedge^2 \mathbf{C}^n)$. Moreover, if $A \in \mathbf{N}_{\mu\mu}^{\mathrm{sym,tr}=0}$, v_A is an eigenvector for the action of $\Delta(x_1, \ldots, x_n)$ with eigenvalue x^μ. Now using the bijection BUR_1, it is easy to see that the characters of both sides in the first identity in Theorem 6.2 are equal. The proof of the second identity is similar. □

REMARK 6.3. The identities of characters in Theorem 6.2 are the Littlewood identities (6) and (7) in [**4**].

$$\prod_{i<j}(1 + x_i x_j) = \sum_{\lambda \text{ threshold}} s_\lambda(x)$$

$$\prod_{i \le j}(1 + x_i x_j) = \sum_{\lambda' \text{ threshold}} s_\lambda(x).$$

7. Representations of Symmetric Groups

If W_1 and W_2 are polynomial representations of $\mathrm{GL}_n(\mathbf{C})$ of degree d, and $T : W_1 \to W_2$ is a homomorphism of representations, then for each μ, $T(W_1(\mu)) \subset W_2(\mu)$. Here $W_i(\mu)$ denotes the μ weight space of W_i (Definition 3.7). The symmetric group S_n may be regarded as a subgroup of $\mathrm{GL}_n(\mathbf{C})$ via permutation matrices. Given a polynomial representation W of $\mathrm{GL}_n(\mathbf{C})$ of *degree* n, let $R_n(W) = W(1, \ldots, 1)$. Clearly, the action of S_n leaves $R_n(W)$ invariant, making it a representation of the symmetric group. Moreover, for a morphism $T : W_1 \to W_2$ of polynomial representations of $\mathrm{GL}_n(\mathbf{C})$, let $R_n(T) : R_n(W_1) \to R_n(W_2)$ denote the linear map obtained by restriction of T to $R_n(W)$. We have a functor:

$$R_n : \mathrm{Rep}^n(\mathrm{GL}_n(\mathbf{C})) \to \mathrm{Rep}(S_n),$$

where $\mathrm{Rep}^n(\mathrm{GL}_n(\mathbf{C}))$ denotes the category of homogeneous polynomial representations of $\mathrm{GL}_n(\mathbf{C})$ of degree n, and $\mathrm{Rep}(S_n)$ denotes the category of complex representations of S_n. Given a partition $\lambda = (\lambda_1, \dots, \lambda_l)$ of n, define:

$$X_\lambda = \{(S_1, \dots, S_l) \mid S_1 \sqcup \dots \sqcup S_l = \{1, \dots, n\}, |S_i| = \lambda_i\}.$$

The action of S_n on $\{1, \dots, n\}$ induces an action on X_λ. This action makes $\mathbf{C}[X_\lambda]$, the space of functions $X_\lambda \to \mathbf{C}$, a representation of S_n. Also, let ϵ denote the sign character of S_n.

Recall a characterization of irreducible complex representations of S_n:

THEOREM 7.1 ([**11**, Appendix A (5.6)], [**12**, Section 4.4]). *For every partition λ of n, there exists a unique representation V_λ of S_n which occurs in both $\mathbf{C}[X_\lambda]$ and $\mathbf{C}[X_{\lambda'}] \otimes \epsilon$. As λ runs over all partition of n, V_λ forms a set of representatives of isomorphism classes of irreducible representations of S_n.*

THEOREM 7.2. *For every partition λ of n, $R_n(W_\lambda) \cong V_\lambda$.*

PROOF. Every submultiset of $\{1, \dots, n\}$ is of the form $I = \{1^{m_1}, \dots, n^{m_n}\}$, where $m_i \geq 0$ denotes the multiplicity of i in I. The size of I is, by definition, the sum $m = m_1 + \dots + m_n$. Then $e_I = \prod_{i=1}^n e_i^{m_i}$ may be regarded as a tensor in $\mathrm{Sym}^m \mathbf{C}^n$. Let $J(n, m)$ denote the set of all submultisets of n of size m. For $S = (S_1, \dots, S_l) \in \prod_{k=1}^l J(n, \lambda_k)$, define

$$e_S = e_{S_1} \otimes \dots \otimes e_{S_l} \in \mathrm{Sym}^\lambda \mathbf{C}^n.$$

Then as S runs over elements of $\prod_{i=k}^l J(n, \lambda_k)$, e_S forms a basis of $\mathrm{Sym}^\lambda \mathbf{C}^n$. The basis vector v_S has weight $\mu = (\mu_1, \dots, \mu_n)$, where μ_i is the sum of the multiplicities of i in $S_1, \dots, S_l$. Thus v_S has weight $(1, \dots, 1)$ if and only if the submultisets $S_1, \dots, S_l$ are actually subsets (each element has multiplicity 0 or 1), and moreover, form a partition of $\{1, \dots, n\}$. In other words, $R_n(\mathrm{Sym}^\lambda \mathbf{C}^n)$ is spanned by $\{v_S \mid S \in X_\lambda\}$. It follows that $R_n(\mathrm{Sym}^\lambda \mathbf{C}^n) \cong \mathbf{C}[X_\lambda]$. Similarly, $R_n(\wedge^{\lambda'} \mathbf{C}^n) \cong \mathbf{C}[X_{\lambda'}] \otimes \epsilon$. Since W_λ occurs in $\mathrm{Sym}^\lambda \mathbf{C}^n$ and also in $\wedge^{\lambda'} \mathbf{C}^n$, it follows that $R_n(W_\lambda)$ occurs in $\mathbf{C}[X_\lambda]$ and $\mathbf{C}[X_{\lambda'}] \otimes \epsilon$. Therefore, by Theorem 7.1, $R_n(W_\lambda)$ would have to be isomorphic to V_λ, unless $R_n(W_\lambda) = 0$. But $\dim(R_n(W_\lambda))$ is the coefficient of $x_1 \cdots x_n$ in $s_\lambda(x_1, \dots, x_n)$, which is the number of tableaux of shape λ and weight $(1, \dots, 1)$ (the *standard* tableaux), which is always positive. □

THEOREM 7.3 (Schur-Weyl duality). *Let S_m act on $\otimes^m \mathbf{C}^n$ by permuting the tensor factors, and $\mathrm{GL}_n(\mathbf{C})$ act on each tensor factor. Then, as a representation of $S_m \times \mathrm{GL}_n(\mathbf{C})$,*

$$\otimes^m \mathbf{C}^n \cong \bigoplus_{\lambda \in \Lambda_n^m} V_\lambda \otimes W_\lambda.$$

PROOF. Apply the functor $R_m \otimes \mathrm{id}$ to $(\mathrm{GL}_m(\mathbf{C}), \mathrm{GL}_n(\mathbf{C}))$ duality (Theorem 4.1). The (1^m)-weight space of $\mathrm{Sym}^m(\mathbf{C}^m \otimes \mathbf{C}^n)$ is spanned by vectors of the form v_A, where each of the m rows of A sums to 1. If the 1 in the ith row occurs in the j_ith column, then map v_A to the vector $e_{j_1} \otimes \dots \otimes e_{j_m} \in \otimes^m \mathbf{C}^n$. This induces an isomorphism $(R_m \otimes \mathrm{id})(\mathbf{C}^m \otimes \mathbf{C}^n) \tilde{\to} \otimes^m \mathbf{C}^n$ of $S_m \times \mathrm{GL}_n(\mathbf{C})$-representations. □

REMARK 7.4. Schur-Weyl duality was used by Schur [**13**] to give a second proof of the classification of irreducible polynomial representations of $\mathrm{GL}_n(\mathbf{C})$. It was popularized by Hermann Weyl [**18**] and is known as Schur-Weyl duality. See also [**6**] and [**12**, Section 6.4].

THEOREM 7.5 (Gelfand models for S_n). *Let $M_{n,k}$ denote the set of all elements $w \in S_n$ such that $w = w^{-1}$ and w has k fixed points. Define a representation of S_n on $\mathbf{C}[M_{n,k}]$ (the space of complex-valued functions on $M_{n,k}$) by:*

$$\rho_1(g)f(w) = (-1)^{i_1(g,w)} f(g^{-1}wg)$$

where $i_1(g,w)$ is the number of indices $i < j$ such that $w(i) = j$ and $g(i) > g(j)$. Then

$$\mathbf{C}[M_{n,k}] = \bigoplus_{\{\lambda \vdash n \mid \text{ with } k \text{ odd columns}\}} V_\lambda.$$

In particular, if $M_n = \coprod M_{n,k}$ is the set of all elements $w \in S_n$ such that $w^2 = \mathrm{id}$, then $\mathbf{C}[M_n]$ is a Gelfand model for S_n.

PROOF. Apply the functor R_n to both sides of the first identity in Theorem 4.9. □

REMARK 7.6. The above Gelfand model for symmetric groups is well-known. For an alternative approach see [**8**, **10**]. The decomposition in Theorem 5.2 also gives a Gelfand model for S_n; the model obtained is the twist of the model in Theorem 7.5 by the sign character. Note that the twist of V_λ by the sign character is $V_{\lambda'}$.

REMARK 7.7 (Relation to induced representations). The representation ρ_1 of S_n on $\mathbf{C}[M_{n,k}]$ in Theorem 7.5 can be viewed as an induced representation. For each positive integer l, let $B_{2l} \subset S_{2l}$ denote the centralizer of the involution $(1,2)(3,4)\cdots(2l-1,2l) \in S_{2l}$. Let e_{2l} denote the restriction of the sign character of S_{2l} to B_{2l}. If $n = 2l + k$, then $B_{2l} \times S_k$ is a subgroup of S_n. We have $\rho_1 = \mathrm{Ind}_{B_{2l}\times S_k}^{S_n} e_{2l} \otimes 1_{S_k}$, where 1_{S_k} denotes the trivial character of S_k. Taking $k = 0$ gives:

$$\mathrm{Ind}_{B_{2l}}^{S_{2l}} e_{2l} = \bigoplus_{\{\lambda \vdash 2l \mid \text{ every part appearing even number of times}\}} V_\lambda.$$

Similarly, applying the functor R_{2l} to both sides of Theorem 5.3 gives:

$$\mathrm{Ind}_{B_{2l}}^{S_{2l}} 1_{B_{2l}} = \bigoplus_{\{\lambda \vdash 2l \mid \text{ with all parts even}\}} V_\lambda.$$

See [**11**, Ch. VII, (2.4)]. For a q-analog of this result, see [**11**, Ch. IV.6, Exercise 6].

Applying the functor R_n to the first identity in Theorem 6.2 gives an interesting multiplicity-free representation of S_n:

THEOREM 7.8. *Let n be an even integer, say $n = 2m$. Given pairs (i,j) and (k,l), let (i',j') and (k',l') denote their sorted versions, so that $i' < j'$ and $k' < l'$. Write $(i,j) < (k,l)$ if $(i',j') < (k',l')$ in lexicographic order. Define a representation of S_n on $\mathbf{C}[M_{n,0}]$ by:*

$$\rho_2(g)f(w) = (-1)^{i_2(g,w)} f(g^{-1}wg),$$

where

$$i_2(g,w) = i_1(g,w) + \#\{(i,j) < (k,l) \mid i < j, k < l, w(i) = j, w(k) = l, \\ (g(i), g(j)) > (g(k), g(l))\}.$$

Then

$$\mathbf{C}[M_{n,0}] = \bigoplus_{\lambda \in \mathrm{TP}(n)} V_\lambda.$$

REMARK 7.9 (Relation to induced representations). The representation ρ_2 of S_{2m} on $M_{2m,0}$ in Theorem 7.8 can be viewed as an induced representation. The group B_{2m} is a semidirect product $(\mathbf{Z}/2\mathbf{Z})^m \rtimes S_m$. The η_m be the character of B_{2m} whose restriction to S_m is the sign character of S_m and whose restriction to $(\mathbf{Z}/2\mathbf{Z})^m$ is the product of the non-trivial characters of each of the m factors. Then Theorem 7.8 can be restated as:

$$\mathrm{Ind}_{B_{2m}}^{S_{2m}} \eta_m = \bigoplus_{\lambda \in \mathrm{TP}(n)} V_\lambda.$$

For a different approach to this result see [**3**, Exercise 45.6].

When R_n is applied to the second identity in Theorem 6.2, the representations obtained are twists by the sign character of the representations in Theorem 7.8.

References

[1] C. Benson and G. Ratcliff, *On multiplicity free actions*, Representations of real and p-adic groups, Lect. Notes Ser. Inst. Math. Sci. Natl. Univ. Singap., vol. 2, Singapore Univ. Press, Singapore, 2004, pp. 221–304, DOI 10.1142/9789812562500_0006. MR2090872

[2] I. N. Bernšteĭn, I. M. Gel′fand, and S. I. Gel′fand, *Models of representations of Lie groups* (Russian), Trudy Sem. Petrovsk. **Vyp. 2** (1976), 3–21. MR0453927

[3] D. Bump, *Lie groups*, Graduate Texts in Mathematics, vol. 225, Springer-Verlag, New York, 2004. MR2062813

[4] W. H. Burge, *Four correspondences between graphs and generalized Young tableaux*, J. Combinatorial Theory Ser. A **17** (1974), 12–30, DOI 10.1016/0097-3165(74)90024-7. MR0351884

[5] W. Fulton, *Young tableaux: With applications to representation theory and geometry*, London Mathematical Society Student Texts, vol. 35, Cambridge University Press, Cambridge, 1997. MR1464693

[6] J. A. Green, *Polynomial representations of* GL_n, Second corrected and augmented edition, Lecture Notes in Mathematics, vol. 830, Springer, Berlin, 2007. With an appendix on Schensted correspondence and Littelmann paths by K. Erdmann, Green and M. Schocker. MR2349209

[7] R. Howe, *Perspectives on invariant theory: Schur duality, multiplicity-free actions and beyond*, The Schur lectures (1992) (Tel Aviv), Israel Math. Conf. Proc., vol. 8, Bar-Ilan Univ., Ramat Gan, 1995, pp. 1–182. MR1321638

[8] N. F. J. Inglis, R. W. Richardson, and J. Saxl, *An explicit model for the complex representations of* S_n, Arch. Math. (Basel) **54** (1990), no. 3, 258–259, DOI 10.1007/BF01188521. MR1037615

[9] D. E. Knuth, *Permutations, matrices, and generalized Young tableaux*, Pacific J. Math. **34** (1970), 709–727. MR0272654

[10] V. Kodiyalam and D.-N. Verma. A natural representation model for symmetric groups. arXiv:math/0402216, 2004.

[11] I. G. Macdonald, *Symmetric functions and Hall polynomials*, 2nd ed., Oxford Mathematical Monographs, The Clarendon Press, Oxford University Press, New York, 1995. MR1354144

[12] A. Prasad, *Representation theory: A combinatorial viewpoint*, Cambridge Studies in Advanced Mathematics, vol. 147, Cambridge University Press, New Delhi, 2015. MR3287258

[13] A. Prasad, An introduction to Schur polynomials. Notes from a course at the *ATM Workshop on Schubert Varieties*, held at The Institute of Mathematical Sciences, Chennai, in November 2017. arXiv:1802.06073, 2018.

[14] A. Sadhukhan. *General Linear Group and Symmetric Group: Commuting Actions and Combinatorics.* M.Sc. thesis, The Institute of Mathematical Sciences (HBNI), Chennai, 2017. `https://www.imsc.res.in/xmlui/handle/123456789/434`

[15] M.-P. Schützenberger, *La correspondance de Robinson* (French), Combinatoire et représentation du groupe symétrique (Actes Table Ronde CNRS, Univ. Louis-Pasteur Strasbourg, Strasbourg, 1976), Springer, Berlin, 1977, pp. 59–113. Lecture Notes in Math., Vol. 579. MR0498826

[16] S. Sundaram. *On the Combinatorics of Representations of* $Sp(2n, \mathbf{C})$. PhD Thesis. Massachusetts Institute of Technology, 1989. Available from `https://dspace.mit.edu/handle/1721.1/15060`.

[17] G. Viennot, *Une forme géométrique de la correspondance de Robinson-Schensted* (French), Combinatoire et représentation du groupe symétrique (Actes Table Ronde CNRS, Univ. Louis-Pasteur Strasbourg, Strasbourg, 1976), Springer, Berlin, 1977, pp. 29–58. Lecture Notes in Math., Vol. 579. MR0470059

[18] H. Weyl, *The Classical Groups. Their Invariants and Representations*, Princeton University Press, Princeton, N.J., 1939. MR0000255

The Institute of Mathematical Sciences (HBNI) Chennai
Email address: `digjoypaul@gmail.com`

The Institute of Mathematical Sciences (HBNI) Chennai
Email address: `amri@imsc.res.in`

University of Maryland
Email address: `arghya.sadhukhan0@gmail.com`

Contemporary Mathematics
Volume **738**, 2019
https://doi.org/10.1090/conm/738/14882

A relationship between Gelfand-Tsetlin bases and Chari-Loktev bases for irreducible finite dimensional representations of special linear Lie algebras

K. N. Raghavan, B. Ravinder, and Sankaran Viswanath

ABSTRACT. We consider two bases for an arbitrary finite dimensional irreducible representation of a complex special linear Lie algebra: the classical Gelfand-Tsetlin basis and the relatively new Chari-Loktev basis. Both are parametrized by the set of (integral Gelfand-Tsetlin) patterns with a fixed bounding sequence determined by the highest weight of the representation. We define the *row-wise dominance* partial order on this set of patterns, and prove that the transition matrix between the two bases is triangular with respect to this partial order. We write down explicit expressions for the diagonal elements of the transition matrix.

1. Introduction

Let $\mathfrak{g} = \mathfrak{sl}_{r+1}$ be the simple Lie algebra of complex matrices of size $(r+1) \times (r+1)$ and trace zero. There are various bases known for (finite dimensional) irreducible representations of $\mathfrak{g}$. Here we consider two of them. The first is the well known basis constructed by Gelfand-Tsetlin in [**2**]. We consider it as reformulated by Molev in [**3**, Theorem 4.3]. The details are recalled in §4 below.

The second basis is of more recent vintage and is obtained as a by-product of the construction by Chari-Loktev [**1**] of bases for local Weyl modules of the current algebra of $\mathfrak{g}$. We consider the construction of Chari-Loktev as reformulated by the present authors in [**4**, Theorem 4.3]. Dominant integral weights of $\mathfrak{g}$ form a natural parametrizing set for local Weyl modules of the current algebra (as they do for irreducible representations of $\mathfrak{g}$). Local Weyl modules admit a natural grading by non-negative integers, and the grade zero slice of a local Weyl module is precisely the corresponding irreducible representation of $\mathfrak{g}$. The Chari-Loktev bases for local Weyl modules are graded. Thus, by

---

2010 *Mathematics Subject Classification.* 17B10 (05E10).

Key words and phrases. Chari-Loktev basis, Gelfand-Tsetlin basis, Upper triangularity, Row-wise dominance order.

The first author is the corresponding author.

KNR and SV acknowledge support under a DAE project grant. BR acknowledges support from DST under the INSPIRE Faculty scheme (DST/INSPIRE/04/2016/001471) and from the Infosys Foundation.

choosing only those elements that have grade zero from the Chari-Loktev basis for a local Weyl module, we obtain a basis for the corresponding irreducible representation of $\mathfrak{g}$. The details are recalled in §5 below.

Both the Gelfand-Tsetlin and the Chari-Loktev bases (for an irreducible representation of $\mathfrak{g}$) are parametrized by (integral Gelfand-Tsetlin) patterns with a fixed bounding sequence determined by the highest weight of the representation. The requisite details about patterns are recalled in §3 below. It is natural to wonder if there is a relationship between the two bases. The goal of the present note is Theorem 4 in §6 below, which shows that the transition matrix between the two bases is triangular with respect to the *row-wise dominance* partial order (defined in the beginning of §6) on their common parametrizing set. Equation (6.2) gives an explicit formula for the diagonal entries of the transition matrix. A simple example illustrating Theorem 4 is given in §7.

2. Notation and terminology

The following notation and terminology remain fixed throughout this note. Let $\mathfrak{g} = \mathfrak{sl}_{r+1}$ be the simple Lie algebra of complex matrices of size $(r+1)\times(r+1)$ and trace zero. Let $\mathfrak{h}$ be the Cartan subalgebra of $\mathfrak{g}$ consisting of its diagonal matrices, and $\mathfrak{b}$ the Borel subalgebra of $\mathfrak{g}$ consisting of its upper triangular matrices. When we speak of roots, weights, dominant weights, etc. in the sequel, it is with respect to this choice of the Cartan and Borel subalgebras.

Throughout, λ denotes a fixed dominant integral weight, $V(\lambda)$ the irreducible representation of $\mathfrak{g}$ with highest weight λ, and v_λ a non-zero element in the λ-weight space of $V(\lambda)$. Note that v_λ is unique up to scaling.

We denote by $\underline{\lambda}$ an arbitrarily fixed member of the equivalence class of $(r+1)$-tuples associated to λ as in §2.1 below. The results below are so formulated that the choice of $\underline{\lambda}$ does not matter.

2.1. Weights and associated tuples. Recall that a *weight* is any linear functional on $\mathfrak{h}$. For i, $1 \leq i \leq r+1$, let ϵ_i denote the weight that maps an element of $\mathfrak{h}$ to its diagonal entry in position (i,i). Evidently, the ϵ_i, $1 \leq i \leq r+1$, span the space of all weights, so any weight can be written as $a_1\epsilon_1 + \cdots + a_{r+1}\epsilon_{r+1}$ for some complex numbers $a_1, \ldots, a_{r+1}$. Given an $(r+1)$-tuple $(a_1, \ldots, a_{r+1})$ of complex numbers, we associate with it the weight $a_1\epsilon_1 + \cdots + a_{r+1}\epsilon_{r+1}$. Since the trace of any element of $\mathfrak{h}$ is zero, two such tuples differ from each other by a scalar multiple of the constant $(r+1)$-tuple $(1, \cdots, 1)$ if and only if the weights associated to them are the same.

A weight $a_1\epsilon_1 + \cdots + a_{r+1}\epsilon_{r+1}$ is *integral* if it can be represented by an $(r+1)$-tuple of integers, or, equivalently, if $a_1 - a_2$, $a_2 - a_3$, $\ldots$, $a_r - a_{r+1}$ are all integers. It is *dominant integral* if $a_1 - a_2$, $a_2 - a_3$, $\ldots$, $a_r - a_{r+1}$ are all non-negative integers.

3. Patterns

Given two non-increasing sequences $\underline{\pi}$: $\pi_1 \geq \ldots \geq \pi_n$ and $\underline{\mu}$: $\mu_1 \geq \ldots \geq \mu_{n-1}$ of real numbers, of lengths n and $n-1$ (for some integer $n > 1$), we say that they *interlace* if $\pi_i \geq \mu_i \geq \pi_{i+1}$ for all i, $1 \leq i < n$.

A pattern $\mathscr{P}$ is a sequence $\underline{\pi}^1, \ldots, \underline{\pi}^{r+1}$ of non-increasing sequences of real numbers, such that:

- $\underline{\pi}^j$ has length j for all j, $1 \leq j \leq r+1$; and
- $\underline{\pi}^{j+1}$ and $\underline{\pi}^j$ interlace for all j, $1 \leq j < r+1$.

We write $\underline{\pi}^j$ as $\pi_1^j \geq \ldots \geq \pi_j^j$, so that the interlacing condition above becomes:

$$\pi_i^j \geq \pi_i^{j-1} \quad \text{and} \quad \pi_i^{j-1} \geq \pi_{i+1}^j \quad \forall\, 1 \leq i < j \leq r+1. \tag{3.1}$$

Let $\mathscr{P}$: $\underline{\pi}^1, \ldots, \underline{\pi}^{r+1}$ be a pattern. Its last sequence $\underline{\pi}^{r+1}$ is called its *bounding sequence*. We say that $\mathscr{P}$ is *integral* if the non-negative real numbers

$$\pi_i^j - \pi_i^{j-1} \text{ and } \pi_i^{j-1} - \pi_{i+1}^j \tag{3.2}$$

are all integers for $1 \leq i < j \leq r+1$.

Observe that the weight associated to the bounding sequence $\underline{\pi}^{r+1}$ of an integral pattern is an integral weight (in the sense of §2.1). In the sequel, we will be interested exclusively in integral patterns with bounding sequence $\underline{\lambda}$, where $\underline{\lambda}$ is fixed as in §2. We denote the set of these patterns by $\mathscr{P}(\underline{\lambda})$.

3.1. Depicting a pattern. A pattern is usually depicted as we illustrate now by means of an example. Let $r = 3$ and $\underline{\lambda} = (8, 6, 3, 1)$. Then the sequence $4;\ 5 \geq 4;\ 7 \geq 5 \geq 2;\ 8 \geq 6 \geq 3 \geq 1$ of sequences is a pattern belonging to $\mathscr{P}(\underline{\lambda})$. It is depicted as follows:

$$\begin{array}{ccccccc} & & & 4 & & & \\ & & 5 & & 4 & & \\ & 7 & & 5 & & 2 & \\ 8 & & 6 & & 3 & & 1 \end{array}$$

Let $\mathscr{P}$: $\underline{\lambda}^1, \ldots, \underline{\lambda}^{r+1} = \underline{\lambda}$ be a pattern in $\mathscr{P}(\underline{\lambda})$. The sequence $\underline{\lambda}^k$ is sometimes referred to as the kth *row* of $\mathscr{P}$. This terminology is justified by the depiction of $\mathscr{P}$. It is also convenient to refer to λ_i^k as the entry in *position* i on row k of $\mathscr{P}$ (where $\underline{\lambda}^k$ is $\lambda_1^k \geq \ldots \geq \lambda_k^k$).

The *weight* of $\mathscr{P}$ is $\mu_1\epsilon_1 + \cdots + \mu_{r+1}\epsilon_{r+1}$, where $\mu_k = (\sum_{i=1}^k \lambda_i^k) - (\sum_{i=1}^{k-1} \lambda_i^{k-1})$. In other words, μ_k is the difference of the sums of the entries in the kth row and the $(k-1)^{\text{st}}$ row of $\mathscr{P}$. For example, the weight of the pattern depicted above is $4\epsilon_1 + 5\epsilon_2 + 5\epsilon_3 + 4\epsilon_4$.

As is easily seen, the weight of a pattern $\mathscr{P}$ in $\mathscr{P}(\underline{\lambda})$ is integral. As is also easily seen, there is a unique pattern in $\mathscr{P}(\underline{\lambda})$ of weight λ. For example, in case $r = 4$ and $\underline{\lambda} = (8, 6, 3, 1)$, this unique pattern is depicted below:

$$\begin{array}{ccccccc} & & & 8 & & & \\ & & 8 & & 6 & & \\ & 8 & & 6 & & 3 & \\ 8 & & 6 & & 3 & & 1 \end{array}$$

4. The Gelfand-Tsetlin (GT) basis for $V(\lambda)$

Theorem 1 below follows Molev's reformulation [**3**, Theorem 4.3] of the original theorem of Gelfand-Tsetlin [**2**].

THEOREM 1. *Let $\mathscr{P}(\underline{\lambda})$ denote the set of integral patterns with bounding sequence $\underline{\lambda}$ (see §3). There exists a basis $\{\xi_{\mathscr{P}}\}$ of $V(\lambda)$ indexed by patterns $\mathscr{P}$ in $\mathscr{P}(\underline{\lambda})$ such that*

- *if $\mathscr{P}$ is the unique pattern in $\mathscr{P}(\underline{\lambda})$ of weight $\underline{\lambda}$, then $\xi_{\mathscr{P}} = v_\lambda$;*
- *for every k, $1 \leq k < r+1$, we have:*

$$(E_{k,k} - E_{k+1,k+1})\,\xi_{\mathscr{P}} = ((\sum_{i=1}^{k} \lambda_i^k - \sum_{i=1}^{k-1} \lambda_i^{k-1}) - (\sum_{i=1}^{k+1} \lambda_i^{k+1} - \sum_{i=1}^{k} \lambda_i^k))\xi_{\mathscr{P}}, \tag{4.1}$$

$$E_{k,k+1}\,\xi_{\mathscr{P}} = -\sum_{i=1}^{k} \frac{(\ell_i^k - \ell_1^{k+1})\cdots(\ell_i^k - \ell_{k+1}^{k+1})}{(\ell_i^k - \ell_1^k)\cdots\wedge\cdots(\ell_i^k - \ell_k^k)}\xi_{\mathscr{P}+\delta_i^k}, \text{ and} \tag{4.2}$$

$$E_{k+1,k}\,\xi_{\mathscr{P}} = \sum_{i=1}^{k} \frac{(\ell_i^k - \ell_1^{k-1})\cdots(\ell_i^k - \ell_{k-1}^{k-1})}{(\ell_i^k - \ell_1^k)\cdots\wedge\cdots(\ell_i^k - \ell_k^k)}\xi_{\mathscr{P}-\delta_i^k}, \tag{4.3}$$

where

- *$E_{i,j}$ denotes the $(r+1)\times(r+1)$ matrix with its only non-zero entry being 1 in position (i,j),*
- *with $\mathscr{P}$ equal to $\underline{\lambda}^1, \ldots, \underline{\lambda}^r, \underline{\lambda}^{r+1}$, we define $\ell_i^k := \lambda_i^k - i + 1$,*
- *$\wedge$ indicates that the vanishing factor $(\ell_i^k - \ell_i^k)$ in the denominator is skipped,*
- *$\mathscr{P} \pm \delta_i^k$ is obtained from $\mathscr{P}$ by replacing λ_i^k by $\lambda_i^k \pm 1$, and*
- *$\xi_{\mathscr{P}\pm\delta_i^k}$ is understood to be 0 if $\mathscr{P} \pm \delta_i^k$ is not a pattern.*

REMARK 2. Since $E_{k,k} - E_{k+1,k+1}$, $E_{k,k+1}$, and $E_{k+1,k}$ generate $\mathfrak{g}$ as a Lie algebra as k ranges over $1 \leq k < r+1$, the equations (4.1)–(4.3) determine completely the action of $\mathfrak{g}$ on $V(\lambda)$. Equation (4.1) is equivalent to saying that $\xi_{\mathscr{P}}$ is a weight vector of weight equal to the weight of $\mathscr{P}$. Thus we obtain as an immediate consequence the following formula for the character of $V(\lambda)$:

$$\text{character of } V(\lambda) = \sum_{\mathscr{P}\in\mathscr{P}(\underline{\lambda})} \exp(\mathrm{wt}(\mathscr{P})). \tag{4.4}$$

5. The Chari-Loktev (CL) basis for $V(\lambda)$

Let $\underline{\lambda}^1, \underline{\lambda}^2, \ldots, \underline{\lambda}^r, \underline{\lambda}^{r+1} = \underline{\lambda}$ be the rows of an integral pattern $\mathscr{P}$ with bounding sequence $\underline{\lambda}$. As in Theorem 1, put $\ell_i^j := \lambda_i^j - i + 1$. Let x_{ij}^-, for $1 \leq i \leq j \leq r$, denote the $(r+1)\times(r+1)$ complex matrix all of whose entries are zero except the one in position $(j+1, i)$ which is 1. (Observe that x_{ij}^- is just another symbol for the matrix that is already denoted by $E_{j+1,i}$ in Theorem 1. Its introduction is prompted by the need to preserve notation of [**4**, Theorem 4.3] which we will be invoking.)

For $1 \leq j \leq r$, let

$$\Theta_{\mathscr{P}}^j := \prod_{i=1}^{j} (x_{ij}^-)^{(\ell_i^{j+1} - \ell_i^j)}$$

where, for an operator T and a non-negative integer n, the symbol $T^{(n)}$ denotes the *divided power* $T^n/n!$. The order of factors in the above product is immaterial since the matrices x_{ij}^- as i varies (but j is fixed) commute with each other.

Finally, let

$$\Theta_{\mathscr{P}} := \Theta_{\mathscr{P}}^1 \cdot \Theta_{\mathscr{P}}^2 \cdot \cdots \cdot \Theta_{\mathscr{P}}^{r-1} \cdot \Theta_{\mathscr{P}}^r \quad \text{and} \quad v_{\mathscr{P}} := \Theta_{\mathscr{P}} v_\lambda. \tag{5.1}$$

Note that the order of the factors does matter in the expression for $\Theta_{\mathscr{P}}$. The following theorem is an immediate consequence of the reformulation by the present authors [**4**, Theorem 4.3] of the original theorem of Chari-Loktev [**1**, Theorem 2.1.3]. We shall call $\Theta_{\mathscr{P}}$ the *Chari-Loktev (or CL)* monomial corresponding to $\mathscr{P}$ and, in view of the theorem, $v_{\mathscr{P}}$ the *Chari-Loktev (or CL)* basis element corresponding to $\mathscr{P}$.

THEOREM 3. *As $\mathscr{P}$ varies over the set $\mathscr{P}(\underline{\lambda})$ of integral patterns with bounding sequence $\underline{\lambda}$, the vectors $v_{\mathscr{P}}$ form a basis of the irreducible representation $V(\lambda)$ of $\mathfrak{g} = \mathfrak{sl}_{r+1}$ with highest weight λ.*

6. Triangular relationship between GT and CL bases of $V(\lambda)$

We state and prove Theorem 4 which is the goal of the present note.

6.1. Statement of the result. On the set $\mathscr{P}(\underline{\lambda})$ of integral patterns with bounding sequence $\underline{\lambda}$, we impose a partial order which we call the *row-wise dominance* order and denote by $\geq$. For $\mathscr{P}$: $\underline{\lambda}^1, \ldots, \underline{\lambda}^r, \underline{\lambda}^{r+1} = \underline{\lambda}$ and $\mathscr{P}'$: $\underline{\lambda}'^1, \ldots, \underline{\lambda}'^r, \underline{\lambda}'^{r+1} = \underline{\lambda}$ in $\mathscr{P}(\underline{\lambda})$, we let:

$$\begin{aligned} \mathscr{P}' \geq \mathscr{P} \quad &\text{if} \quad \underline{\lambda}'^j \geq \underline{\lambda}^j \quad \text{for all } 1 \leq j \leq r+1, \text{ where} \\ \underline{\lambda}'^j \geq \underline{\lambda}^j \quad &\text{if} \quad \lambda_1'^j + \cdots + \lambda_i'^j \geq \lambda_1^j + \cdots + \lambda_i^j \text{ for all } 1 \leq i \leq j. \end{aligned}$$

THEOREM 4. *For any pattern $\mathscr{P}$: $\underline{\lambda}^1, \ldots, \underline{\lambda}^r, \underline{\lambda}^{r+1} = \underline{\lambda}$ in $\mathscr{P}(\underline{\lambda})$, in the expression for the corresponding Chari-Loktev basis element $v_{\mathscr{P}}$ of $V(\lambda)$ (see Theorem 3) as a linear combination of the Gelfand-Tsetlin basis $\{\xi_{\mathscr{P}'}\}_{\mathscr{P}' \in \mathscr{P}(\underline{\lambda})}$ (see Theorem 1), only those $\xi_{\mathscr{P}'}$ with $\mathscr{P}' \geq \mathscr{P}$ appear. In other words, we can write:*

$$v_{\mathscr{P}} = \sum_{\mathscr{P}' \geq \mathscr{P}} c_{\mathscr{P}}^{\mathscr{P}'} \xi_{\mathscr{P}'} \tag{6.1}$$

where $c_{\mathscr{P}}^{\mathscr{P}'}$ are entries of the transition matrix between the two bases. The diagonal element $c_{\mathscr{P}}^{\mathscr{P}}$ in the transition matrix is given by:

$$c_{\mathscr{P}}^{\mathscr{P}} = \prod_{1 \leq i < j \leq r+1} \prod_{p=i}^{j-1} \frac{(\ell_i^{j-1} - \ell_p^{j-1})!}{(\ell_i^j - \ell_p^{j-1})!} \tag{6.2}$$

where, as in Theorem 1, $\ell_i^j := \lambda_i^j - i + 1$.

6.2. Proof of Theorem 4. Proceed by induction on the length of $\mathscr{P}$, where length is defined by

$$\text{length}\,\mathscr{P} := \sum_{1 \leq i < j \leq r+1} (\lambda_i^j - \lambda_i^{j-1}) = \sum_{1 \leq i < j \leq r+1} (\ell_i^j - \ell_i^{j-1}). \tag{6.3}$$

If the length is 0, then $\mathscr{P}$ is the unique one of weight $\underline{\lambda}$, the only pattern $\mathscr{P}'$ such that $\mathscr{P}' \geq \mathscr{P}$ is $\mathscr{P}$ itself, $v_{\mathscr{P}} = \xi_{\mathscr{P}} = v_\lambda$, and $c_{\mathscr{P}}^{\mathscr{P}} = 1$, and we are done.

Let now the length of $\mathscr{P}$ be at least one. Let $j_0 := \min\{j \geq 2 \,|\, \lambda_i^j \neq \lambda_i^{j-1}$ for some $1 \leq i \leq j\}$ and $i_0 := \min\{i \,|\, \lambda_i^{j_0} \neq \lambda_i^{j_0-1}\}$. Let $\tilde{\mathscr{P}}$: $\tilde{\underline{\lambda}}^1, \ldots, \tilde{\underline{\lambda}}^r, \tilde{\underline{\lambda}}^{r+1} = \tilde{\underline{\lambda}}$ be obtained from $\mathscr{P}$ by replacing $\lambda_{i_0}^j$ by $\lambda_{i_0}^j + 1$ for $i_0 \leq j < j_0$ (the other entries stay the same). See Figure 6.1.

It is easily seen that

$$\tilde{\mathscr{P}} \text{ is a pattern}, \quad \tilde{\mathscr{P}} \geq \mathscr{P}, \quad \text{and} \quad \text{length}\,\tilde{\mathscr{P}} = \text{length}\,\mathscr{P} - 1.$$

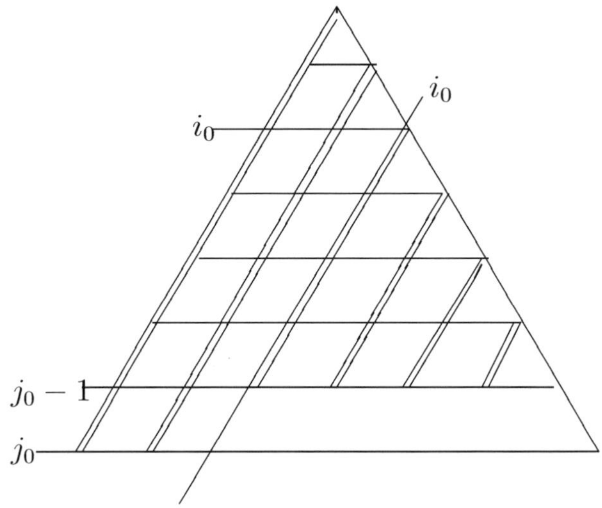

FIGURE 6.1. The figure shows the first j_0 rows of the pattern $\mathscr{P}$. The Southwest–Northeast double lines indicate equalities, while the lone single Southwest–Northeast line between rows j_0 and $j_0 - 1$ indicates $\gneq$.

It is also easily seen from the definitions in §5 that the Chari-Loktev monomials of $\mathscr{P}$ and $\tilde{\mathscr{P}}$, and consequently also the corresponding basis elements, are related thus:

$$\Theta_{\mathscr{P}} = \frac{1}{\ell_{i_0}^{j_0} - \ell_{i_0}^{j_0-1}} x^-_{i_0,j_0-1} \Theta_{\tilde{\mathscr{P}}} \quad \text{and} \quad v_{\mathscr{P}} = \frac{1}{\ell_{i_0}^{j_0} - \ell_{i_0}^{j_0-1}} E_{j_0,i_0} v_{\tilde{\mathscr{P}}}. \tag{6.4}$$

By the induction hypothesis, we have:

$$v_{\tilde{\mathscr{P}}} = \sum_{\mathscr{P}' \geq \tilde{\mathscr{P}}} c_{\tilde{\mathscr{P}}}^{\mathscr{P}'} \xi_{\mathscr{P}'} \quad \text{and} \quad c_{\tilde{\mathscr{P}}}^{\tilde{\mathscr{P}}} = \prod_{1 \le i < j \le r+1} \prod_{p=i}^{j-1} \frac{(\tilde{\ell}_i^{j-1} - \tilde{\ell}_p^{j-1})!}{(\tilde{\ell}_i^{j} - \tilde{\ell}_p^{j-1})!} \tag{6.5}$$

where $\tilde{\ell}_i^j := \tilde{\lambda}_i^j - i + 1$. It is convenient to rewrite the expression above for $c_{\tilde{\mathscr{P}}}^{\tilde{\mathscr{P}}}$ in terms of the numbers $\ell_i^j = \lambda_i^j - i + 1$ for $\mathscr{P}$. We claim that

$$c_{\tilde{\mathscr{P}}}^{\tilde{\mathscr{P}}} = (\ell_{i_0}^{j_0} - \ell_{i_0}^{j_0-1})\Big(\prod_{p=i_0+1}^{j_0-1} (\ell_{i_0}^{j_0-1} - \ell_p^{j_0-1} + 1)\Big) \left(\prod_{1 \le i < j \le r+1} \prod_{p=i}^{j-1} \frac{(\ell_i^{j-1} - \ell_p^{j-1})!}{(\ell_i^{j} - \ell_p^{j-1})!} \right). \tag{6.6}$$

To prove the claim, we first observe the following:

$$\prod_{p=i}^{j-1} \frac{(\tilde{\ell}_i^{j-1} - \tilde{\ell}_p^{j-1})!}{(\tilde{\ell}_i^{j} - \tilde{\ell}_p^{j-1})!} = \prod_{p=i}^{j-1} \frac{(\ell_i^{j-1} - \ell_p^{j-1})!}{(\ell_i^{j} - \ell_p^{j-1})!} \quad \text{for } (j,i) \neq (j_0, i_0). \tag{6.7}$$

Indeed, for $j \geq j_0$, we have $\underline{\lambda}^j = \tilde{\underline{\lambda}}^j$, so (6.7) holds for $j > j_0$; it also holds for $i > i_0$ since $\ell_i^j = \tilde{\ell}_i^j$ in that case; finally, if either $j < j_0$ or $j = j_0$ and $i < i_0$, then $\ell_i^j = \ell_i^{j-1}$ and $\tilde{\ell}_i^j = \tilde{\ell}_i^{j-1}$, so the product on either side of (6.7) equals 1.

Continuing with the proof of the claim (6.6), we calculate the left hand side of (6.7) in the case $(j,i)=(j_0,i_0)$:

$$\prod_{p=i_0}^{j_0-1}\frac{(\tilde\ell_{i_0}^{j_0-1}-\tilde\ell_p^{j_0-1})!}{(\tilde\ell_{i_0}^{j_0}-\tilde\ell_p^{j_0-1})!}=\frac{1}{(\tilde\ell_{i_0}^{j_0}-\tilde\ell_{i_0}^{j_0-1})!}\times\prod_{p=i_0+1}^{j_0-1}\frac{(\tilde\ell_{i_0}^{j_0-1}-\tilde\ell_p^{j_0-1})!}{(\tilde\ell_{i_0}^{j_0}-\tilde\ell_p^{j_0-1})!}$$

$$=\frac{1}{(\ell_{i_0}^{j_0}-\ell_{i_0}^{j_0-1}-1)!}\times\prod_{p=i_0+1}^{j_0-1}\frac{(\ell_{i_0}^{j_0-1}-\ell_p^{j_0-1}+1)!}{(\ell_{i_0}^{j_0}-\ell_p^{j_0-1})!}$$

$$=\frac{(\ell_{i_0}^{j_0}-\ell_{i_0}^{j_0-1})}{(\ell_{i_0}^{j_0}-\ell_{i_0}^{j_0-1})!}\times\left(\prod_{p=i_0+1}^{j_0-1}(\ell_{i_0}^{j_0-1}-\ell_p^{j_0-1}+1)\right)\times\left(\prod_{p=i_0+1}^{j_0-1}\frac{(\ell_{i_0}^{j_0-1}-\ell_p^{j_0-1})!}{(\ell_{i_0}^{j_0}-\ell_p^{j_0-1})!}\right)$$

$$=(\ell_{i_0}^{j_0}-\ell_{i_0}^{j_0-1})\times\left(\prod_{p=i_0+1}^{j_0-1}(\ell_{i_0}^{j_0-1}-\ell_p^{j_0-1}+1)\right)\times\left(\prod_{p=i_0}^{j_0-1}\frac{(\ell_{i_0}^{j_0-1}-\ell_p^{j_0-1})!}{(\ell_{i_0}^{j_0}-\ell_p^{j_0-1})!}\right)$$

where, to justify the last equality, we observe that the $(\ell_{i_0}^{j_0}-\ell_{i_0}^{j_0-1})!$ in the denominator of the first factor is incorporated into the last factor by changing the lower range of p from i_0+1 to i_0. This calculation along with (6.7) proves the claim (6.6).

We now prove (6.1) of the theorem. Substituting the expression for $v_{\tilde{\mathscr{P}}}$ in (6.5) into the expression for $v_{\mathscr{P}}$ in (6.4), we get

$$v_{\mathscr{P}}=\frac{1}{\ell_{i_0}^{j_0}-\ell_{i_0}^{j_0-1}}\sum_{\mathscr{P}'\geq\tilde{\mathscr{P}}}c_{\tilde{\mathscr{P}}}^{\mathscr{P}'}E_{j_0,i_0}\xi_{\mathscr{P}'}. \tag{6.8}$$

Observe that

$$E_{j_0,i_0}=[E_{j_0,j_0-1}[E_{j_0-1,j_0-2},\cdots[E_{i_0+2,i_0+1},E_{i_0+1,i_0}]\cdots]]. \tag{6.9}$$

So the operator E_{j_0,i_0} on $V(\lambda)$ is a linear combination with coefficients ±1 of the operators

$$E_{\sigma(j_0),\sigma(j_0)-1}E_{\sigma(j_0-1),\sigma(j_0-1)-1}\cdots E_{\sigma(i_0+1),\sigma(i_0+1)-1} \tag{6.10}$$

as σ varies over permutations of the set $\{i_0+1,\ldots,j_0\}$, with 1 being the coefficient of the operator

$$E_{j_0,j_0-1}E_{j_0-1,j_0-2}\cdots E_{i_0+2,i_0+1}E_{i_0+1,i_0}.$$

Now, (6.1) follows from formula (4.3) for the action of $E_{k+1,k}$ in Theorem 1 and item (2) of the following lemma.

LEMMA 5. *Let $\mathscr{P}'$ be any pattern in $\mathscr{P}(\underline{\lambda})$ such that $\mathscr{P}'\geq\tilde{\mathscr{P}}$. Let $\mathscr{P}''$ be any pattern obtained from $\mathscr{P}'$ by decreasing by 1 exactly one entry in each of the rows i_0, …, j_0-1. Then*

(1) $\mathscr{P}''\in\mathscr{P}(\underline{\lambda})$.
(2) $\mathscr{P}''\geq\mathscr{P}$.
(3) $\mathscr{P}''=\mathscr{P}$ *only if* $\mathscr{P}'=\tilde{\mathscr{P}}$.

(4) *The coefficient of $\xi_{\mathscr{P}}$ in $E_{j_0,i_0}\xi_{\tilde{\mathscr{P}}}$ equals that of $\xi_{\mathscr{P}}$ in $E_{j_0,j_0-1}E_{j_0-1,j_0-2}\cdots E_{i_0+1,i_0}\xi_{\tilde{\mathscr{P}}}$, and consequently:*

(6.11)
$$\text{coefficient of } \xi_{\mathscr{P}} \text{ in } E_{j_0,i_0}\xi_{\tilde{\mathscr{P}}} = \prod_{k=i_0}^{j_0-1} \text{coefficient of } \xi_{\tilde{\mathscr{P}}(k+1)} \text{ in } E_{k+1,k}\,\xi_{\tilde{\mathscr{P}}(k)}$$

where for k in the range $i_0 \leq k \leq j_0$, $\tilde{\mathscr{P}}(k)$ denotes the pattern whose initial rows up to but not including the kth are those of $\mathscr{P}$ and the remaining ones are those of $\tilde{\mathscr{P}}$ (so that $\tilde{\mathscr{P}}(i_0) = \tilde{\mathscr{P}}$ and $\tilde{\mathscr{P}}(j_0) = \mathscr{P}$).

PROOF: Since $j_0 - 1 < r+1$, it is clear that $\mathscr{P}''$ shares its $(r+1)^{\text{st}}$ row with $\mathscr{P}'$ and so (1) is clear. Let the rows of $\mathscr{P}'$ be $\underline{\lambda}'^1, \ldots, \underline{\lambda}'^r, \underline{\lambda}'^{r+1} = \underline{\lambda}$ and those of $\mathscr{P}''$ be $\underline{\lambda}''^1, \ldots, \underline{\lambda}''^r, \underline{\lambda}''^{r+1} = \underline{\lambda}$.

Towards the proof of (2), first observe that for j such that $1 \leq j < i_0$ or $j_0 \leq j \leq r+1$, we have

$$\underline{\lambda}''^j = \underline{\lambda}'^j \geq \tilde{\underline{\lambda}}^j = \underline{\lambda}^j.$$

It remains to show that for j such that $i_0 \leq j < j_0$ and k such that $1 \leq k \leq j$, it holds that

$$\lambda_1''^j + \cdots + \lambda_k''^j \geq \lambda_1^j + \cdots + \lambda_k^j. \tag{6.12}$$

For such j, let r_j denote the position in row j in which $\mathscr{P}''$ differs from $\mathscr{P}'$.

We consider three cases: $k < r_j$, $i_0 \leq k$, and $r_j \leq k < i_0$ (this last may not occur if $i_0 \leq r_j$). In the first case we have

$$\lambda_1''^j + \cdots + \lambda_k''^j = \lambda_1'^j + \cdots + \lambda_k'^j \geq \tilde{\lambda}_1^j + \cdots + \tilde{\lambda}_k^j \geq \lambda_1^j + \cdots + \lambda_k^j$$

and in the second case

$$\lambda_1''^j + \cdots + \lambda_k''^j \geq \lambda_1'^j + \cdots + \lambda_k'^j - 1 \geq \tilde{\lambda}_1^j + \cdots + \tilde{\lambda}_k^j - 1 = \lambda_1^j + \cdots + \lambda_k^j.$$

In the third case, using the fact that $\mathscr{P}''$ is a pattern, we have:

$$\lambda_1''^j + \cdots + \lambda_k''^j \geq \lambda_1''^k + \cdots + \lambda_k''^k = \lambda_1'^k + \cdots + \lambda_k'^k \geq \tilde{\lambda}_1^k + \cdots + \tilde{\lambda}_k^k = \lambda_1^k + \cdots + \lambda_k^k = \lambda_1^j + \cdots + \lambda_k^j$$

which completes the proof of (2).

Let us now prove (3). Since $\mathscr{P}''$ is given to be $\mathscr{P}$, it follows that $\mathscr{P}'$ differs from $\mathscr{P}$ in exactly one position on each of the rows $i_0, \ldots, j_0-1$ and furthermore the difference in each case is only that the entry of $\mathscr{P}'$ is precisely one more than the corresponding entry of $\mathscr{P}$. Thus, as far as other rows are concerned, that is, for j not in the range $i_0 \leq j < j$, we have $\underline{\lambda}'^j = \underline{\lambda}^j = \tilde{\underline{\lambda}}^j$.

Now let j be fixed such that $i_0 \leq j < j$. We need to show $\underline{\lambda}'^j = \tilde{\underline{\lambda}}^j$, or, in more detail, $\lambda_i'^j = \tilde{\lambda}_i^j$ for all i, $1 \leq i \leq j$. Suppose that we have proved this for i in the range $1 \leq i < i_0$. Then, since

$$\lambda_1'^j + \cdots + \lambda_{i_0-1}'^j + \lambda_{i_0}'^j \geq \tilde{\lambda}_1^j + \cdots + \tilde{\lambda}_{i_0-1}^j + \lambda_{i_0}^j$$

(this equation holds because $\mathscr{P}' \geq \tilde{\mathscr{P}}$), it would follow that $\lambda_{i_0}'^j \geq \tilde{\lambda}_{i_0}^j = \lambda_{i_0}^j + 1$. But given how little $\mathscr{P}'$ is allowed to differ from $\mathscr{P}$, it would follow that equality holds here, that is, $\lambda_{i_0}'^j = \tilde{\lambda}_{i_0}^j = \lambda_{i_0}^j + 1$. Moreover it would follow for the same reason that $\lambda_i'^j = \lambda_i^j = \tilde{\lambda}_i^j$ for i in the range $i > i_0$ as well, and we would be done with the proof of (3).

Thus it only remains to prove that $\lambda'^j_i = \tilde{\lambda}^j_i$ for i in the range $1 \leq i < i_0$, which we proceed to do by induction on i. The induction hypothesis is that $\lambda'^j_s = \tilde{\lambda}^j_s$ for $1 \leq s < i$ (which is vacuously true in the base case $i = 1$). Since

$$\lambda'^j_1 + \cdots + \lambda'^j_{i-1} + \lambda'^j_i \geq \tilde{\lambda}^j_1 + \cdots + \tilde{\lambda}^j_{i-1} + \lambda^j_i$$

(this equation holds because $\mathscr{P}' \geq \tilde{\mathscr{P}}$), it follows that $\lambda'^j_i \geq \tilde{\lambda}^j_i$. Using this (as the fourth (in)equality in the string of (in)equalities below), we have:

$$\lambda^{j_0}_i = \lambda''^{j_0}_i = \lambda'^{j_0}_i \geq \lambda'^j_i \geq \tilde{\lambda}^j_i = \lambda^j_i = \lambda^{j_0}_i \qquad \text{so that } \lambda'^j_i = \tilde{\lambda}^j_i (= \lambda^j_i = \lambda^{j_0}_i)$$

where, the first equality is justified because $\mathscr{P}''$ is given to be equal to $\mathscr{P}$; the second because $\mathscr{P}''$ and $\mathscr{P}'$ have the same j_0^{th} row (among others); the third because $\mathscr{P}'$ is a pattern; the fifth because $\tilde{\mathscr{P}}$ differs from $\mathscr{P}$ only in positions i_0 in rows $i_0, \ldots, j_0 - 1$; and, finally, the sixth by the choice of i_0 and j_0. This finishes the proof of (3).

Turning to (4), given how E_{j_0,i_0} is a linear combination of operators as in (6.10) and the formula (4.3) for the action of $E_{k+1,k}$ on a Gelfand-Tsetlin basis vector, it suffices to prove the following:

> Suppose that $\mathscr{Q}_{i_0}, \ldots, \mathscr{Q}_{j_0}$ be a sequence of patterns (indexed by k in the range $i_0 \leq k \leq j_0$) and σ a permutation of the set $\{i_0, \ldots, j_0 - 1\}$ such that $\mathscr{Q}_{i_0} = \tilde{\mathscr{P}}$, $\mathscr{Q}_{j_0} = \mathscr{P}$, and, for each k, $i_0 \leq k < j_0$, the pattern $\mathscr{Q}_{k+1}$ is obtained from $\mathscr{Q}_k$ by making only one change, namely, decreasing the entry in position i_0 in the row $\sigma(k)$. Then σ is the identity permutation.

By way of contradiction, suppose that σ is not the identity. Let k be least such that $\sigma(k) \neq k$. Then $\sigma(k) > k$. Compare the entries in position i_0 on rows $\sigma(k)$ and k in the pattern $\mathscr{Q}_{k+1}$. The former is 1 less than the latter, which contradicts the assumption that $\mathscr{Q}_{k+1}$ is a pattern. This finishes the proof of (4) and so also of the lemma. □

We now turn to the proof of (6.2) of Theorem 4. Combining item (3) of Lemma 5 with (6.8), we obtain the following:

$$c^{\mathscr{P}}_{\mathscr{P}} = \frac{c^{\tilde{\mathscr{P}}}_{\tilde{\mathscr{P}}}}{\ell^{j_0}_{i_0} - \ell^{j_0-1}_{i_0}} \times \text{the coefficient of } \xi_{\mathscr{P}} \text{ in } E_{j_0,i_0}\xi_{\tilde{\mathscr{P}}}. \tag{6.13}$$

Substituting for $c^{\tilde{\mathscr{P}}}_{\tilde{\mathscr{P}}}$ its value from (6.6), we see that it suffices to prove the following:

LEMMA 6. *The coefficient of $\xi_{\mathscr{P}}$ in $E_{j_0,i_0}\xi_{\tilde{\mathscr{P}}}$ equals*

$$\prod_{p=i_0+1}^{j_0-1} \frac{1}{(\ell^{j_0-1}_{i_0} - \ell^{j_0-1}_{p} + 1)}. \tag{6.14}$$

PROOF: Item (4) of Lemma 5 says that the required coefficient equals

$$\prod_{p=i_0}^{j_0-1} \text{coefficient of } \xi_{\tilde{\mathscr{P}}(p+1)} \text{ in } E_{p+1,p}\,\xi_{\tilde{\mathscr{P}}(p)}.$$

From (4.3) in Theorem 1, we see that, for p in the range $i_0 \leq p < j_0$, the coefficient of $\xi_{\tilde{\mathscr{P}}(p+1)}$ in $E_{p+1,p}\,\xi_{\tilde{\mathscr{P}}(p)}$ is

$$\frac{\prod_{q=1}^{p-1}(\alpha_{i_0}+1-\alpha_q)}{\prod_{q=1}^{i_0-1}(\alpha_{i_0}+1-\alpha_q)\times\prod_{q=i_0+1}^{p}(\alpha_{i_0}+1-\alpha_q)} = \frac{1}{(\alpha_{i_0}+1-\alpha_p)} = \frac{1}{(\ell_{i_0}^{j_0-1}-\ell_p^{j_0-1}+1)}$$

where we have denoted by α_q the value $\ell_q^q = \ell_q^{q+1} = \ldots = \ell_q^{j_0-1}$ (for q in the range $1 \leq q < j_0$). This finishes the proof of Lemma 6 and so also that of Theorem 4. □

7. An example illustrating Theorem 4

Let $\mathfrak{g} = \mathfrak{sl}_3$ (that is, $r = 2$). Let $\lambda = 4\epsilon_1 + 2\epsilon_2$ and $\underline{\lambda} = 4 \geq 2 \geq 0$. In $\mathscr{P}(\underline{\lambda})$ there are three patterns of weight 0 and here they are listed in row-wise dominance order:

$$\begin{array}{ccccc} & & 2 & & \\ & 4 & & 0 & \\ 4 & & 2 & & 0 \end{array} \quad \geq \quad \begin{array}{ccccc} & & 2 & & \\ & 3 & & 1 & \\ 4 & & 2 & & 0 \end{array} \quad \geq \quad \begin{array}{ccccc} & & 2 & & \\ & 2 & & 2 & \\ 4 & & 2 & & 0 \end{array}$$

Let us denote them by $\mathscr{P}_1$, $\mathscr{P}_2$, and $\mathscr{P}_3$ respectively. Letting ξ_1, ξ_2, and ξ_3 be the corresponding GT basis elements (of the zero weight space) of $V(\lambda)$ and v_1, v_2, and v_3 the corresponding Chari-Loktev basis elements, we have:

$$\begin{pmatrix} v_1 \\ v_2 \\ v_3 \end{pmatrix} = \begin{pmatrix} 1/4 & 0 & 0 \\ -1/4 & 1/4 & 0 \\ 1/24 & -1/8 & 1/12 \end{pmatrix} \begin{pmatrix} \xi_1 \\ \xi_2 \\ \xi_3 \end{pmatrix}$$

References

[1] V. Chari, S. Loktev, *Weyl, Demazure and fusion modules for the current algebra of* $\mathfrak{sl}_{r+1}$, *Adv. Math.* **207** (2006), no. 2, 928–960, arXiv:math/0502165.

[2] I. M. Gelfand, M. L. Tsetlin, *Finite-dimensional representations of the group of unimodular matrices*, Doklady Akad. Nauk SSSR **71**, (1950), 825–828 (Russian). MR0035774; English transl. in: I. M. Gelfand, "Collected papers". Vol II, Berlin: Springer-Verlag 1988, pp. 653–656.

[3] A. I. Molev, *Gelfand-Tsetlin bases for classical Lie algebras*, Handbook of algebra. Vol. 4, Handb. Algebr., vol. 4, Elsevier/North-Holland, Amsterdam, 2006, pp. 109–170, DOI 10.1016/S1570-7954(06)80006-9. MR2523420

[4] K. N. Raghavan, B. Ravinder, and S. Viswanath, *On Chari-Loktev bases for local Weyl modules in type A*, J. Combin. Theory Ser. A **154** (2018), 77–113, DOI 10.1016/j.jcta.2017.08.011. MR3718061

The Institute of Mathematical Sciences, HBNI, CIT campus, Taramani, Chennai 600113, India

Email address: knr@imsc.res.in

Chennai Mathematical Institute, H1 Sipcot IT Park, Kelambakkam, Siruseri, Chennai 603103, INDIA

Email address: bravinder@cmi.ac.in

The Institute of Mathematical Sciences, HBNI, CIT campus, Taramani, Chennai 600113, India

Email address: svis@imsc.res.in

Contemporary Mathematics
Volume **738**, 2019
https://doi.org/10.1090/conm/738/14883

Algebra of generic matrices is not coherent

Lance Small and Efim Zelmanov

Abstract. We prove that the algebra generated by m generic $n \times n$ matrices over an infinite field, $m > 1$, $n > 1$, (i) is not finitely presented, (ii) is not coherent.

1. Introduction

The concepts of coherence and finite presentation of algebras are similar: each refers to presentations of modules or algebras that involve only a finite number of relations. In this note we investigate these properties for the algebra of generic matrices over a field (for definitions and background we refer to [**2**] and [**4**]). It turns out that generic matrices are neither finitely presented nor coherent. The proofs of these results are, not surprisingly, intertwined.

All vector spaces are over an infinite field F. Consider the algebra of polynomials $F\left[x_{ij}^{(k)} \;\middle|\; 1 \leq i,j \leq n;\ 1 \leq k \leq m\right]$ in mn^2 variables and $n \times n$ matrices

$$X_k = \left(x_{ij}^{(k)}\right)_{1 \leq i,j \leq n}, \quad 1 \leq k \leq m$$

over it. The F-algebra $G(m,n)$ generated by $X_1, \ldots, X_m$ is called the algebra of generic matrices (see [**4**]).

Consider the free algebra $\mathrm{Fr}(m)$ in m free generators $x_1, \ldots, x_m$ and the homomorphism $\mathrm{Fr}(m) \xrightarrow{\varphi} G(m,n)$, $x_k \mapsto X_k$, $1 \leq k \leq m$. Let $I = \ker(\varphi)$. The celebrated theorem of A. Kemer ([**1**], [**2**]) says that the ideal I is finitely generated as a T-ideal. We prove that I is not finitely generated as an ideal, which means that the algebra $G(m,n)$ is not finitely presented.

Theorem 1. *The algebra $G(m,n)$ is not finitely presented.*

Recall that an algebra A is called (right) coherent if every finitely generated right ideal of A is finitely presented as a right A-module. Every affine commutative algebra is coherent. For more information about coherent algebras see [**3**].

Our next result is

Theorem 2. *The algebra $G(m,n)$ is not coherent.*

2010 *Mathematics Subject Classification.* Primary 16R10.
Key words and phrases. generic matrices, finitely presented algebra, coherent algebra.
The second author is the corresponding author.

2. Proofs

We begin with

LEMMA 1. *The matrix algebra $M_n(F)$ does not satisfy a homogeneous identity $f = 0$ in two variables x, y such that the degree of f in y is $\leq n-1$.*

PROOF. Suppose that

$$f(x,y) = \sum_{i_1+\ldots+i_k=d} \alpha_{i_1\cdots i_k} x^{i_1} y x^{i_2} \cdots y x^{i_k} = 0$$

is an identity of $M_n(F)$, $k \leq n$. Linearizing in y we get

$$\sum \alpha_{i_1\cdots i_k} \left(\sum_{\sigma\in\mathcal{S}_{k-1}} x^{i_1} y_{\sigma(1)} x^{i_2} \cdots y_{\sigma(k-1)} x^{i_k} \right) = 0.$$

Choose $y_1 = e_{12}$, $y_2 = e_{23}, \ldots, y_{k-1} = e_{k-1,k}$, $x = \mathrm{diag}(\lambda_1, \lambda_2, \ldots, \lambda_k, 0, 0, \ldots, 0)$, where the e_{ij} are matrix units, $\lambda_i \in F$. Then

$$\sum_{\sigma\in\mathcal{S}_{k-1}} x^{i_1} y_{\sigma(1)} x^{i_2} \cdots y_{\sigma(k-1)} x^{i_k} = x^{i_1} y_1 x^{i_2} \cdots y_{k-1} x^{i_k} = e_{1k}(\lambda_1^{i_1} \lambda_2^{i_2} \cdots \lambda_k^{i_k}).$$

It follows that $\sum \alpha_{i_1\cdots i_k} \lambda_1^{i_1} \cdots \lambda_k^{i_k} = 0$ for arbitrary $\lambda_1, \ldots, \lambda_k \in F$. Since the field F is infinite we conclude that all $\alpha_{i_1\cdots i_k} = 0$, which completes the proof of the lemma. □

The following lemma is well-known (see [**2**]).

LEMMA 2. *For arbitrary $i_1, \ldots, i_{n+1} \in \mathbb{Z}_{\geq 0}$ the matrix algebra $M_n(F)$ satisfies the identity*

$$f_{i_1,\ldots,i_{n+1}}(x,y) = \sum_{\sigma\in\mathcal{S}_{n+1}} (-1)^{\mathrm{sgn}(\sigma)} x^{i_{\sigma(1)}} y x^{i_{\sigma(2)}} y \cdots y x^{i_{\sigma(n+1)}} = 0.$$

PROOF OF THEOREM 1. If $m \geq 2$ and the algebra $G(m,n)$ is finitely presented then the algebra $G(2,n)$ is finitely presented as well. Therefore we will prove the theorem only for the algebra $G(2,n) = \langle X_1, X_2 \rangle$ where X_1, X_2 are generic $n \times n$ matrices.

Consider the free algebra $\mathrm{Fr}(2)$ on free generators x, y and the homomorphism $x \xrightarrow{\varphi} X_1$, $y \xrightarrow{\varphi} X_2$; $I = \ker(\varphi)$. Suppose that the ideal I is generated (as an ideal) by elements $h_1(x,y), \ldots, h_s(x,y)$. Without loss of generality we will assume $h_1, \ldots, h_s$ to be homogeneous in x and y. By Lemma 1 the degree of each $h_i(x,y)$ in y is $\geq n$, $1 \leq i \leq s$.

There exists a sufficiently large $r \geq 1$ such that $h_1(x,y), \ldots, h_s(x,y)$ lie in the span of monomials of type $x^{i_1} y x^{i_2} y \cdots y x^{i_{t+1}}$, where $0 \leq i_1, \ldots, i_{t+1} \leq r$.

Choose $i_1 = r+1$, $i_2 = r+2, \ldots, i_{n+1} = r+n+1$. By Lemma 2 the element $f_{i_1,\ldots,i_{n+1}}(x,y)$ lives in I, hence

$$f_{i_1,\ldots,i_{n+1}} = \sum_{i,j} \alpha_{ij} u_{ij} h_i v_{ij}, \tag{1}$$

where $\alpha_{ij} \in F$; u_{ij}, v_{ij} are homogeneous elements of $\mathrm{Fr}(2)$. By Lemma 1 every h_i has degree ≥ 2 in y. Hence every monomial on the right hand side of (1) has a subword $yx^k y$, $k \leq r$. It is easy to see that no monomial on the left hand side of (1) has such a subword. This completes the proof of Theorem 1. □

PROOF OF THEOREM 2. Again without loss of generality we assume that $m = 2$ and consider the algebra $G = G(2,n)$. It will be convenient to denote the generic matrices as x and y, so $G = \langle x, y\rangle$.

Consider the right ideal $\rho = xG + yG$. Let $uG + vG$ be the free G-module on the free generators u, v. Consider the homomorphism $\psi : uG+vG \longrightarrow \rho$, $\psi(u) = x$, $\psi(v) = y$. Let $M = \ker(\psi)$. We will show that M is not finitely generated as a G-module.

Suppose that M is generated by elements $ua_i + vb_i$, $1 \le i \le s$, where a_i, b_i are homogeneous elements in x, y. For a homogeneous element a let $\deg_x(a)$, $\deg_y(a)$ denote its degrees in x, y respectively.

By Lemma 1 $\deg_y(b_i) \ge n-1$, $\deg_y(a_i) = \deg_y(b_i) + 1 \ge n$. Let

$$N > \max(\deg_x(a_1), \dots, \deg_x(a_s),\ n-1).$$

Choose $i_1 = 0$, $i_2 = 1$, $i_3 = 2, \dots, i_n = n-1$, $i_{n+1} = N$. By Lemma 2 we have $f_{i_1,\dots,i_{n+1}}(x,y) = 0$. Let's collect all monomials of $f_{i_1,\dots,i_{n+1}}$ that start with x and all monomials that start with y,

$$f_{i_1,\dots,i_{n+1}}(x,y) = xg(x,y) + yh(x,y).$$

It is easy to see that

$$h(x,y) = \sum_{\sigma\in H} (-1)^{\mathrm{sgn}(\sigma)} x^{i_{\sigma(2)}} y x^{i_{\sigma(3)}} \cdots y x^{i_{\sigma(n+1)}},$$

where H is the stabilizer of 1 in $\mathcal{S}_{n+1}$.

We will determine $g(x,y)$ modulo the left ideal Gx. Modulo Gx we need to consider only summands that correspond to permutations $\sigma \in \mathcal{S}_{n+1}$ such that $\sigma(n+1) = 1$. Hence,

$$g(x,y) = \sum_{\substack{\sigma\in\mathcal{S}_{n+1}\\ \sigma(n+1)=1}} (-1)^{\mathrm{sgn}(\sigma)} x^{i_{\sigma(1)}-1} y x^{i_{\sigma(2)}} y \cdots y x^{i_{\sigma(n)}} y \quad \mathrm{mod}\ Gx.$$

LEMMA 3.

$$\sum_{\substack{\sigma\in\mathcal{S}_{n+1}\\ \sigma(n+1)=1}} (-1)^{\mathrm{sgn}(\sigma)} x^{i_{\sigma(1)}-1} y x^{i_{\sigma(2)}} y \cdots y x^{i_{\sigma(n)}} y \notin Gx.$$

PROOF. If the expression on the left hand side lies in Gx, then, linearizing in y, we get that for arbitrary elements $y_1, \dots, y_n \in G$

$$(2) \qquad \sum_{\tau\in\mathcal{S}_n} \sum_{\substack{\sigma\in\mathcal{S}_{n+1}\\ \sigma(n+1)=1}} (-1)^{\mathrm{sgn}(\sigma)} x^{i_{\sigma(1)}-1} y_{\tau(1)} x^{i_{\sigma(2)}} y_{\tau(2)} \cdots x^{i_{\sigma(n)}} y_{\tau(n)} \in Gx.$$

Choose $y_1 = e_{12}$, $y_2 = e_{23}, \dots, y_{n-1} = e_{n-1,n}$, $y_n = e_{n,1}$, $x = \mathrm{diag}(0, \lambda_2, \dots, \lambda_n)$, where $\lambda_i \in F$, $2 \le i \le n$. We have $xe_{11} = 0$ and $y_ie_{11} = 0$ for all $i < n$. Multiplying both sides of the inclusion (2) by e_{11} on the right we get

$$\sum_{\substack{\sigma\in\mathcal{S}_{n+1}\\ \sigma(n+1)=1}} (-1)^{\mathrm{sgn}(\sigma)} x^{i_{\sigma(1)}-1} e_{12} x^{i_{\sigma(2)}} e_{23} \cdots e_{n-1,n} x^{i_{\sigma(n)}} e_{n1} = 0.$$

The element $x^{i_{\sigma(1)}-1}e_{12}$ may be different from 0 only if $i_{\sigma(1)} = 1$, so $\sigma(1) = 2$. Now we get

$$\sum_{\substack{\sigma\in S_{n+1}\\ \sigma(n+1)=1\\ \sigma(1)=2}} (-1)^{\mathrm{sgn}(\sigma)} e_{12}x^{i_{\sigma(2)}}e_{23}\cdots e_{n-1,n}x^{i_{\sigma(n)}}e_{n1}$$

$$= e_{11}\left(\sum_{\substack{\sigma(n+1)=1\\ \sigma(1)=2}} (-1)^{\mathrm{sgn}(\sigma)}\lambda_2^{i_{\sigma(2)}}\lambda_3^{i_{\sigma(3)}}\cdots\lambda_n^{i_{\sigma(n)}}\right) = 0.$$

Clearly, there is a choice of $\lambda_i \in F$, $2 \le i \le n$, that makes this equality false. This completes the proof of the lemma. □

Now let's continue the proof of Theorem 2. Since $xg(x,y) + yh(x,y) = 0$ it follows that $ug(x,y) + vh(x,y) \in M$. Hence

$$ug(x,y) + vh(x,y) = \sum_{i=1}^{s} (ua_i + vb_i)c_i, \quad c_i \in G.$$

In particular,

$$g(x,y) = \sum_{i=1}^{s} a_i c_i.$$

From $\deg_y g(x,y) = n$ and $\deg_y a_i \ge n$, $1 \le i \le s$, we conclude that all elements c_i are powers of x. Moreover, $\deg_x g(x,y) \ge N > \max_i\{\deg_x(a_i)\}$ implies that all the elements c_i are nontrivial powers of x; hence $g(x,y) \in Gx$ which contradicts Lemma 3. This completes the proof of Theorem 2. □

Acknowledgements

The second author gratefully acknowledges the support by the NSF grant DMS 160 1920.

References

[1] A. R. Kemer, *Identities of finitely generated algebras over an infinite field* (Russian), Izv. Akad. Nauk SSSR Ser. Mat. **54** (1990), no. 4, 726–753, DOI 10.1070/IM1991v037n01ABEH002053; English transl., Math. USSR-Izv. **37** (1991), no. 1, 69–96. MR1073084

[2] A. R. Kemer, *Ideals of identities of associative algebras*, Translations of Mathematical Monographs, vol. 87, American Mathematical Society, Providence, RI, 1991. Translated from the Russian by C. W. Kohls. MR1108620

[3] T. Y. Lam, *Lectures on modules and rings*, Graduate Texts in Mathematics, vol. 189, Springer-Verlag, New York, 1999. MR1653294

[4] C. Procesi, *Rings with polynomial identities*, Marcel Dekker, Inc., New York, 1973. Pure and Applied Mathematics, 17. MR0366968

Department of Mathematics, University of California San Diego, 9500 Gilman Drive, La Jolla, CA 92093-0112, USA

Email address: lwsmall@math.ucsd.edu

Department of Mathematics, University of California San Diego, 9500 Gilman Drive, La Jolla, CA 92093-0112, USA

Email address: ezelmano@math.ucsd.edu

Contemporary Mathematics
Volume **738**, 2019
https://doi.org/10.1090/conm/738/14884

Partial endomorphisms of almost self injective modules

Surjeet Singh

ABSTRACT. Let M be a quasi-injective module and S be its ring of endomorphisms. It is well known that the radical J of S is the set of all those endomorphisms of M which have large kernels, and S/J is a von Neumann regular ring. Such a result does not hold for almost self-injective modules. Let N be an almost self-injective module and G its ring of endomorphisms. By enlarging the ring of endomorphisms of N, it is shown that a generalization of the above result for quasi-injective modules can be given for N. In addition an artinian serial ring R with no non-trivial central idempotent is studied. A characterization for R to be almost right self-injective in terms of its Kuppisch series is given, which can help in creating certain types of almost self-injective rings.

Introduction

For any two modules M and N over a ring R, the concept of M being almost N-injective was introduced by Baba [**3**]. This concept and its dual, the concept of almost relative projectivity were initially also studied by Harada [**5**–**7**]. Recently these concepts have also been studied by Al-Ahmadi and Jain [**2**], and by Singh [**8, 9**]. If a module M is almost M-injective, then M is said to be *almost self-injective*. Any quasi-injective module is almost self-injective and any almost self-injective module M is such that any homomorphism f from a submdule L of M into M with large kernel can be extended to an endomorphism of M, *i.e.* M is essentially self-injective. Let M be an almost self-injective module. Any homomorphism f from a submodule L of M into M is called a *partial endomorphism* of M. A partial endomorphism $f : L \to M$ is said to be *maximal*, if there does not exist a partial endomorphism $g : L_1 \to M$ which extends f and $L < L_1 \leqslant M$. Let T be the set of all maximal partial endomorphisms of M. Then T contains the set S of all endomorphisms of M. Clearly $T = S$ if and only if M is quasi-injective. Any two members f and g of T are said to be *equivalent*, if they are equal on an essential submodule of M. Then the set $\overline{T}$ of equivalence classes in T is made into a ring, which is proved to be von Neumann regular (Theorem 2.5). In section 3, artinian serial ring are studied. Let R be an artinian serial ring having no non-trivial central idempotent. A characterization in terms of the Kuppisch series of R, under which

2010 *Mathematics Subject Classification.* Primary 16D50; secondary 16E50.

Key words and phrases. Quasi-injective modules, uniform modules, von Neumann regular rings, serial rings, Kuppisch series, essentially relative injectivity, almost relative injectivity.

R is almost right self-injective, is given in Theorem 3.3, which can be used in constructing certain types of almost self-injective rings.

1. Preliminaries

All rings considered here are with unity and all modules are unital right modules. Let M_R, N_R be any two modules. Then $L \leqslant M$ ($L < M$) denotes that L is a submodule (respectively proper submodule) of M. If a module M has finite composition length, then $d(M)$ denotes its *composition length.* Symbol $L \subset_e M$ denote that L is an essential submodule of M. Any homomorphism $f : L \to M$, $L \leqslant N$ is called a *partial homomorphism* from N to M. A partial homomorphism $f : L \to M$, $L \leqslant N$ is said to be *maximal,* if there does not exist a homomorphism $f_1 : L_1 \to M$ which extends f and $L < L_1 \leqslant N$. By definition, any homomorphism $f : N \to M$ is a maximal partial homomorphism from N to M. The module M is said to be *almost N-injective,* if for any homomorphism $f : L \to M$, $L < N$, either f extends from N to M or there exists a decompositiom $N = N_1 \oplus N_2$, $N_1 \neq 0$ and a homomorphism $h : M \to N_1$ such that $hf(x) = \pi(x)$ for any $x \in L$, where $\pi : N \to N_1$ is projection with kernel N_2. The following is observed in [**8**, Proposition 1.2].

LEMMA 1.1. *Let M_R, N_R be any two modules. Then M is almost N-injective if and only if for any maximal partial homomorphism f from N to M, either dom(f) $= N$ or there exists a decomposition $M = N_1 \oplus N_2$, $N_1 \neq 0$ and a homomorphism $h : M \to N_1$ such that $hf(x) = \pi(x)$ for any $x \in L$, where $\pi : M \to N_1$ is projection with kernel N_2.*

DEFINITION 1.2. Let M_R, N_R be any two modules. Then M is said to be *essentially N-injective* if any homomorphism $f : L \to M$, $L \leqslant N$ with $\ker f \subset_e L$, can be extended from N to M.

LEMMA 1.3. *Let M_R, N_R be any two modules.*

(i) M is essentially N-injective if and only if for any homomorphism $f : L \to M$, $L \subset_e N$ with $\ker f \subset_e L$, can be extended from N to M.

(ii) If M is almost N-injective, then M is essentially N-injective.

Proof. (i) Let the given condition hold and $f : L \to M$, $L \leqslant N$ be a homomorphism with $\ker f \subset_e L$. Let C be a complement of L in N. Define homomorphism $g : L \oplus C \to M$ which extends f and is zero on C. Then $\ker g \subset_e C \oplus L$. By the hypothesis, g can be extended to a homomorphism $h : N \to M$. Clearly, h extends f. Hence M is essentially N-injective.

(ii) Let M be almost N-injective. Let $f : L \to M$, $L \subset_e N$ be a homomorphism with $\ker f \subset_e L$. Suppose f cannot be extended from N to M. Then there exists a decomposition $N = N_1 \oplus N_2$ with $N_1 \neq 0$ and a homomorphism $h : M \to N_1$ such that $hf(x) = \pi(x)$ for any $x \in L$, where $\pi : N \to N_1$ is projection with kernel N_2. Then $L \cap N_1 \neq 0$ and f restricted to $L \cap N_1$ is monic, which is a contradiction. Hence M is essentially N-injective. □

The following is a consequence of [**8**, Corollary 1.7].

LEMMA 1.4. *Let M, N be any two uniform modules over a ring R. Then M is almost N-injective if and only if for any homomorphism $\sigma : A \to M$, $A < N$, either it extends from N to M or any of its maximal extension $\eta : L \to M$, $A \leq L < N$ is monic and $\eta(L) = M$.*

PROPOSITION 1.5 (See [**2**, Remark 17]]). *Let* $\{U_i : 1 \leq i \leq n\}$ *be a finite family of uniform R-modules, and* $M = \oplus \sum_{i=1}^{n} U_i$. *Then M is almost self-injective if and only if* U_i, U_j *are almost relative injective for any* i, j.

For basic terminology used in this paper one may refer to [**1**,**4**].

2. Partial Homomorphisms

Following generalizes [**9**, Proposition 1.11(ii)] and has similar proof.

PROPOSITION 2.1. *Let Module* M_R, N_R *be any two modules such that M is essentially N-injective.*

(i) Let $f_1 : L_1 \to M$, $L_1 \leqslant N$, $f_2 : L_2 \to M$, $L_2 \leqslant N$ *be any two homomorphisms such that* f_1 *is a maximal partial homomorphism from N to M . Let* $\mu : L_1 \cap L_2 \to M$, $\mu(x) = f_1(x) - f_2(x)$ *and* $g : L \to M$, $L \leqslant N$ *be a maximal extension of* μ. *Then* $L \cap L_2 = L_1 \cap L_2$

(ii) If f_2 *is also a maximal partial homomorphism, then* $L \cap L_1 = L_1 \cap L_2 = L \cap L_2$.

(iii) Let $f_1 : L_1 \to M$, $L_1 \leqslant N$, $f_2 : L_2 \to M$, $L_2 \leqslant N$ *be any two maximal partial homomorphisms from N to M such that for some* $K \subset_e L_1 \cap L_2$, $f_1(x) = f_2(x)$, *whenever* $x \in K$. *Then* $L_1 = L_2$.

Proof.(i) Let $K = L \cap L_2$. Then $K \cap L_1 = L \cap L_2 \cap L_1 = L_1 \cap L_2$. For any $x \in K$, both $g(x)$, $f_2(x)$ are defined, so is $\lambda(x) = g(x) + f_2(x)$. For $y \in K \cap L_1$, $\lambda(y) = \mu(y) + f_2(y) = f_1(y)$. We get $\beta : K + L_1 \to M$, $\beta(x + y) = \lambda(x) + f_1(y)$, $x \in K$, $y \in L_1$. As β is an extension of f_1, the maximality of f_1 gives $K \subseteq L_1$, $L \cap L_2 \subseteq L_1 \cap L_2$. Hence $L \cap L_2 = L_1 \cap L_2$.

(ii) If f_2 is maximal, then by (i), $L \cap L_1 = L_2 \cap L_1$. Hence $L \cap L_1 = L_1 \cap L_2 = L \cap L_2$.

(iii) Let $\mu : L_1 \cap L_2 \to M$ as in (i). As $K \subseteq \ker \mu$, and M is essentially N-injective, we have extension $g : N \to M$ of μ. Then, as in (i), $L = N$. By (ii), $L_1 = L_2$. □

Let M, N, P be any three modules over a ring R. Let $T(N, M)$ be set of all maximal partial homomorphisms from N to M. Observe that the domain of any $f \in T(N, M)$, $dom(f)$ is an essential submodule of N , and $Hom_R(N, M) \subseteq T(N, M)$. Also $Hom_R(N, M) = T(N, M)$ if and only if M is N-injective. For any $f, g \in T(N, M)$. Define $f \sim g$, whenever there exists a $K \subset_e dom(f) \cap dom(g)$ such that $f(x) = g(x)$ for any $x \in K$. This is a equivalence relation in $T(N, M)$. Let $\overline{T}(N, M)$ be the set of equivalence classes under the relation "$\sim$" in $T(N, M)$, and for any $f \in T(N, M)$, $[f]$ denote the equivalence class of f under this equivalence relation. Let $\Delta(N, M)$ be set of all those $f \in T(N, M)$, for which $\ker f \subset_e N$. Then Proposition (2.1) gives the following.

LEMMA 2.2. *Let* M_R, N_R *be two modules such that M is essentially N-injective.*
(i) For two members f_1, f_2 *of an* $[f] \in \overline{T}(N, M)$, $dom(f_1) = dom(f_2)$.
(ii) $\Delta(N, M) \subseteq Hom(N, M)$ *and* $[0] = \Delta(N.M)$. □

LEMMA 2.3. *Let P, M, N be any three modules over a ring R such that M is essentially N-injective and P is essentially M-injective and essentially N-injective.*

(i) For any f, $g \in T(N, M)$, *let* $h \in T(N, M)$ *be a maximal extension of the mapping* $\lambda : dom(f) \cap dom(g) \to M$, $\lambda(x) = f(x) + g(x)$, $x \in dom(f) \cap dom(g)$.

Then $[f]+[g]=[h]$ *defines a binary composition in* $\overline{T}(N,M)$*, under which* $\overline{T}(N,M)$ *is an abelian group.*

(ii) For any $f \in T(N,M)$*,* $g \in T(M,P)$*, let* $L_1 = \{x \in dom(f) : f(x) \in dom(g)\}$ *and* $h \in T(N.P)$ *be such that it extends the mapping* $\mu : L_1 \to P$*,* $\mu(x) = g(f(x))$*,* $x \in L_1$*. Then* $[g][f] = [h]$ *is well defined.*

(iii) For any f_1*,* $f_2 \in T(N,M)$*,* g_1*,* $g_2 \in T(M,P)$*,* $([g_1]+[g_2])[f_1] = [g_1][f_1] + [g_2][f_1]$ *and* $[g_1]([f_1]+[f_2]) = [g_1][f_1] + [g_1][f_2]$*.*

Proof. (i). Observe that for any $\lambda \in T(N,M)$, $dom(\lambda) \subset_e N$. Let $f_1 \in [f]$, $g_1 \in [g]$. Let $K = dom(f) \cap dom(g)$. Then $K \subset_e N$, and $f(x)+g(x)$ is defined for any $x \in K$. We get a $\lambda \in T(N,M)$ such that $\lambda(x) = f(x)+g(x)$ for any $x \in K$. We also have $\lambda' \in T(N,M)$ such that $\lambda'(x) = f_1(x)+g_1(x)$ for any $x \in K' = dom(f_1) \cap dom(g_1)$. By Proposition (2.1), $dom(f) = dom(f_1)$, $dom(g) = dom(g_1)$. Therefore $K = K'$, and by definition of the equivalence relation there exists $K_1 \subset_e dom(f) \cap dom(f_1)$ such that $f_1(x) = f(x)$ for any $x \in K_1$. Similarly, we get $K_2 \subset_e dom(g) \cap dom(g_1)$ such that $g_1(x) = g(x)$ for any $x \in K_2$. Now $K_3 = K \cap K_1 \cap K_2 \subset_e N$ and for any $x \in K \cap K_1 \cap K_2$, $\lambda(x) = f(x)+g(x) = f_1(x)+g_1(x) = \lambda'(x)$. So that $[\lambda] = [\lambda']$. Thus we define $[f]+[g] = [\lambda]$ and see that it is well defined. Given $f \in T(N,M)$, $f \sim 0$ if and only if $\ker f \subset_e N$ and $dom(f) = dom(0) = N$. Which proves that $[0] = \Delta(N,M) = \{f \in Hom(N,M) : \ker f \subset_e N\}$. This makes $\overline{T}(N,M)$ an abelian group in which $0 = \Delta(N,M)$.

(ii) For any $K \subset_e N$, $L \subset_e M$, $f^{-1}(L) \cap K \subset_e N$. So, $L_1 = f^{-1}(dom(g))$ is such that $L_1 \subset_e dom(f)$ and $L_1 \subset_e dom(h) \cap dom(f) \subset_e N$ and $h(x) = g(f(x))$ for any $x \in L_1$. Let $f_1 \in [f]$, $g_1 \in [g]$ and $[g_1][f_1] = [h_1]$ such that for any $x \in f_1^{-1}(dom(g_1)$, $h_1(x) = g_1(f_1(x))$. Set $L_1' = f_1^{-1}(dom(g_1)$, then $L_1' \subset_e dom(h_1) \cap dom(f_1) \subset_e N$. We get $K_1 \subset_e dom(f)$, $K_2 \subset_e dom(g)$ such that $f(x) = f_1(x)$ for any $x \in K_1$, and $g(x) = g_1(x)$ for any $x \in K_2$. Now $H = f_1^{-1}(K_2) \cap K_1 \subset_e dom(f_1) = dom(f)$. Let $x \in H$. Then $f(x) = f_1(x) \in K_2$. Therefore $g(f(x) = g(f_1(x) = g_1(f_1(x))$. Therefore $[h] = [h_1]$. Hence $[g][f]$ is well defined.

(iii) It now follows that for any $f \in T(N,M)$, $g \in T(M,P)$, $h \in T(N,P)$, $[g][f] = [h]$ if and only if there exists $K \subset_e f^{-1}(dom(g)) \cap dom(h))$ such that $h(x)$ $g(f(x))$ for any $x \in K$. Also for any λ, λ_1, $\lambda_2 \in T(N,M)$, $[\lambda_1]+[\lambda_2] = [\lambda]$ if and only if there exists $K \subset_e dom(\lambda_1) \cap dom(\lambda_2) \cap dom(\lambda)$ such that $\lambda_1(x)+\lambda_2(x) = \lambda(x)$ for any $x \in K$. Set $L = f_1^{-1}(dom(g_1) \cap dom(g_2))$. Then $L \subset_e N$ and $(g_1+g_2)j_1(x) = g_1(f_1(x)) + g_2(f_1(x))$ for any $x \in L$. From which it follows that $([g_1]+[g_2])[f_1] = [\mu]$, where $\mu \in T(N,P)$ is such that there exists $K_1 \subset_e L \cap dom(\mu)$ such that $\mu(x) = g_1(f_1(x)) + g_2(f_1(x))$ for any $x \in K_1$. At the same time, we get h_1, $h_2 \in T(N,M)$ such that $[g_1][f_1] = [h_1]$, $[g_2][f_1] = [h_2]$. Then there exists $H_1 \subset_e L \cap dom(h_1) \cap dom(h_2)$ such that $h_1(x) = g_1(f_1(x))$, $g_2(f_1(x)) = h_2(x)$, whenever $x \in H_1$. From this it follows that $[\mu] = [h_1]+[h_2] = [g_1][f_1] + [g_2][f_1]$. Hence $([g_1]+[g_2])[f_1] = [g_1][f_1] + [g_2][f_1]$. Set $H = f_1^{-1}(dom(g_1) \cap f_2^{-1}(dom(g_1))$. Then $H \subset_e N$, and for any $x \in H$, $g_1(f_1(x)+f_2(x)) = g_1(f_1(x)) + g_1(f_2(x))$ is defined. Using this, as above, it follows that $([g_1]([f_1]+[f_2]) = [g_1][f_1]+[g_1][f_2]$. □

For any M_R, set $T(M) = T(M,M)$, $\Delta(M) = \Delta(M,M)$. For any module M_R, $\Delta(M)$ is an ideal of $End(M_R)$ and for $\sigma \in \Delta(M)$, $I-\sigma$ is monic. The following is immediate from the above lemma.

LEMMA 2.4. *Let* M_R *be essentially self-injective. Then* $\Delta(M)$ *is an ideal of* $End(M_R)$ *and* $\overline{T}(M) = \overline{T}(M,M)$ *is a ring.*

THEOREM 2.5. *Let M_R, N_R be two essentially self-injective modules such that they are also relatively essentially injective. Then for any $f \in \overline{T}(N,M)$, there exists $\mu \in T(M,N)$ such that $[f][\mu][f] = [f]$.*

(i) $\overline{T}(M)$ is a von Neumann regular ring.

(ii) If M is non-singular, then $T(M)$ is von Neumann regular.

Proof. Let $f \in T(N,M)$ such that $f \notin \Delta(N,M)$. Let $L = dom(f)$, $K = \ker f$ and A be a complement of K in L. Then $A \cong f(A) \subseteq M$ and $A \neq 0$. Let C be a complement of A containing $\ker f$. Let f_1 be in $T(N,M)$ such that $f|A = f_1|A$ and f_1 is zero on C. Then f and f_1 are equal on the essential submodule $A \oplus \ker f$, therefore $[f] = [f_1]$ and by (2.1), $dom(f) = dom(f_1)$. Set $B = A \oplus C$. Then $B \subseteq dom(f)$. For any $b = a + c \in B$, where $a \in A$, $c \in C$, $f(b) = f(a) + f(c)$. Suppose $f(b) = 0$. Then $f(a+c) = 0$, $a + c \in \ker f$, $a \in A \cap C = 0$, $a = 0$, $b = c \in \ker f$. Thus if $\theta = f|B$, then $\ker \theta = \ker f$. Let $f(a) = f(c)$ for some $a \in A$, $c \in C$. Then $a - c \in \ker f \subseteq C$, $a \in A \cap C = 0$, $f(c) = 0$. This proves that $f(A) \cap f(C) = 0$. Let K' be a complement of $f(A)$ in M containing $f(C)$.

Set $L_1 = f(A) \oplus K' \subseteq M$. Define $\mu : f(A) \oplus K' \to N$, $\mu(f(a) + k) = a$, where $a \in A$, $k \in K'$. We have maximal extension of μ, say $\lambda : L_2 \to N$, $L_2 \leqslant M$. For $x = a + c \in A \oplus C$, $\lambda f(a+c) = \lambda(f(a) + f(c)) = \lambda f(a) = a$, which gives $f\lambda f(a+c) = f(a)$. In particular, if $c \in \ker f$, then $f\lambda f(a+c) = f(a+c)$. Thus $[f\lambda f] = [f]$. Hence $[f][\mu][f] = [f]$.

By taking $M = N$, we immediately get that $\overline{T}(M)$ is a von Neumann regular ring. Finally if M is nonsingular, then $\Delta(M) = 0$, and that gives (ii). □

3. Artinian Serial Rings

Let R be an artinian serial ring [**4**] with no proper central idempotent. Let $e_1, e_2, \ldots, e_n$ be a maximal orthogonal set of non-isomorphic indecomposable idempotents in R. Now each e_iR is a finite length uniserial module, and by re-indexing, we have the following. For any $i < n$, $e_iJ \neq 0$, and is a homomorphic image of $e_{i+1}R$, and if $e_nJ \neq 0$, then it is a homomorphic image of e_1R [**4**, Chap 25]. In this case, we have the following:

(i) For any $i < n$, $d(e_iR) \leq d(e_{i+1}R) + 1$,

(ii) If $e_nJ \neq 0$, then $d(e_nR) \leq d(e_1R) + 1$.

The sequence $e_1R, e_2R, \ldots, e_nR$ is called a *Kuppisch series* of R. Henceforth R is an artinian serial ring having no proper central idempotent. Let $d(e_iR) = m_i$. For $i < n$, $m_i \leqslant m_{i+1} + 1$. If $m_n > 1$, then $m_n \leq m_1 + 1$. Given any two uniserial modules A, B over an artinian serial ring R, B is A-injective, whenever $d(A) \leq d(B)$. From which it follows that that given any two uniserial R-modules A, B any non-zero homomorphism $f : L \to B$, $0 < L < A$ can be extended from A to B if and only if $d(\frac{A}{L}) \leqslant d(\frac{B}{f(L)})$. Any uniserial R-module is quasi-injective and quasi-projective. Given two simple modules X_R, Y_R, Y is called a *successor* of X, if there exists a uniserial module A_R of length 2 such that $soc(A) \cong Y$ and $\frac{A}{soc(A)} \cong X$; in this case X is called a *predecessor* of Y. Over an artinian serial ring a simple modules cannot have more than one successor and more than one predecessor. For basic results on serial rings one may see [**10**] and in particular for artinian serial rings one may see Faith [**4**]. Following is well known.

LEMMA 3.1. *(i) For $i < j \leq n$, $m_i \leq m_j + j - i$.*

(ii) If $m_n > 1$, then for $i < j \leq n$, $m_j \leq m_i + n - j + i$.

(iii) if $m_n = 1$, then no two composition factors of an e_iR are isomorphic, and $m_i \leq n - i + 1$ for any i, and for $i < j$, $Hom(e_iR, e_jR) = 0$

(iv) For any uniserial module A_R, and any two non-negative integers k, l, if $\frac{AJ^k}{AJ^{k+1}}$, $\frac{AJ^l}{AJ^{l+1}}$ are non-zero, then they are isomorphic if and only if $k \equiv 1 (\mathrm{mod}\, n)$.

(i), (ii) and (iii) are immediate from above, and (iv) is known as the periodicity property of R.

In case $m_n > 1$, the members of the above Kuppisch series can be cyclically rotated, say e_2R, e_3R, ... , e_nR, e_1R is also Kuppisch series of R.

By using the periodicity property, one gets the following:

LEMMA 3.2. *For $i < j \leq n$.*

(i) If $m_i \leq j - i$, then e_iR has no composition factor isomorphic to $\frac{e_jR}{e_jJ}$. If $m_i > j - i$, $k = j - i$ is smallest positive integer such that $\frac{e_jR}{e_jJ} \cong \frac{e_iJ^k}{e_iJ^{k+1}}$.

(ii) If $m_j \leq n + i - j$, then e_jR has no composition factor isomorphic to $\frac{e_iR}{e_iJ}$. If $m_j > n + i - j$, then $k = n + i - j$ is the smallest positive integer such that $\frac{e_iR}{e_iJ} \cong \frac{e_jJ^k}{e_jJ^{k+1}}$; in this case $m_n > 1$.

We now prove the following theorem.

THEOREM 3.3. *Let R be an artinian serial ring having no non-trivial central idempotent. Let e_1R, e_2R, ... , e_nR be a Kuppisch series of R and $m_i = d(e_iR)$ for any $1 \leq i \leq n$. Then R_R is almost self-injective if and only if it one of the following holds for any pair of positive integers $i < j \leq n$.*

(i) $m_i = m_j + (j - i)$ or $m_j = m_i + n + (i - j)$, whenever $soc(e_iR) \cong soc(e_jR)$.

(ii) $m_i + i - j < m_j \leq n + i - j$, $m_j + n + i - j < m_i \leq j - i$ and $soc(e_iR) \ncong soc(e_jR)$.

(iii) The smallest length $A < e_iR$ such that $soc(\frac{e_iR}{A}) \cong soc(e_jR)$ exists and is non-zero, $d(A) = m_i - m_j + n + i - j$ and $m_j + n + i - j < m_i \leq j - i$.

(iv) The smallest length $B < e_jR$ such that $soc(\frac{e_jR}{B}) \cong soc(e_iR)$ exists and is non-zero, $d(B) = m_j - m_i + j - i$ and $m_i + i - j < m_j \leq n + i - j$.

(v) The smallest length $A < e_iR$ such that $soc(\frac{e_iR}{A}) \cong soc(e_jR)$ exists and is non-zero, $d(A) = m_i - m_j + n + i - j$, smallest length $B < e_jR$ such that $soc(\frac{e_jR}{B}) \cong soc(e_iR)$ exists and is non-zero, and $d(B) = m_j - m_i + j - i$.

Now every uniform module over an artinian serial ring is quasi-injective, therefore, by (1.5), we need to prove that for any positive integers $i < j \leq n$, e_iR and e_jR are relatively almost injective. Two cases arise according as $soc(e_iR) \cong soc(e_jR\}$ or $soc(e_iR) \ncong soc(e_jR)$. We prove the result through a series of lemmas.

LEMMA 3.4. *Let $i < j \leqslant n$ and $soc(e_iR) \cong soc(e_jR)$. Then the following hold.*

(i) If e_jR is almost e_iR-injective, then either $m_i = m_j + (j - i)$ or $m_j = m_i + n + (i - j)$.

(ii) If e_iR is almost e_jR-injective, then either $m_i = m_j + (j - i)$ or $m_j = m_i + n + (i - j)$.

(iii) e_iR and e_jR are relatively almost self-injective if and only if $m_i = m_j + (j - i)$ or $m_j = m_i + n + (i - j)$.

Proof. By (3.1), $m_i \leq m_j + j - i$. Now e_iJ is a homomorphic image of $e_{i+1}R$, repeated use of this gives e_iJ^{j-i} is a homomorphic image of e_jR. In particular, if $e_iJ^{j-i} \neq 0$, then $\frac{e_iJ^{j-i}}{e_iJ^{j-i+1}} \cong \frac{e_jR}{e_jJ}$.

Suppose e_jR is almost e_iR-injective. By the hypothesis, we have an isomorphism $\sigma : soc(e_iR) \to soc(e_jR)$.Suppose σ does not extend from e_iR to e_jR. By (1.4), σ extends to an an isomorphism $\eta : L \to e_jR$, $L < e_iR$. Then $d(L) = m_j$. By (3.1)(i), $m_i - d(L) \leq j - i$, *i.e.* $d(\frac{e_iR}{L}) \leq j - i$. By (3.2), $L = e_iJ^{j-i}$, which gives $m_i = m_j + j - i$.

Suppose σ extends to a monomorphism η from e_iR into e_jR. The periodicity shows that $\eta(e_iR) \subseteq e_jJ^{n-(j-i)}$, therefore $m_i \leq m_j - n + j - i$, $m_i + n + (i - j) \leq m_j$. If $m_n = 1$, it follows from (3.1) that $m_j \leq n - j + 1$, so $m_i + n + i - j \leq n - j + 1$, $m_i + i \leq 1 - j$, which is a contradiction. Hence $m_n > 1$. Then starting from e_jR and using periodicity, we get $m_j \leq m_i + n + (i - j)$. Hence $m_j = m_i + n + (i - j)$. This proves (i).

(ii) Suppose e_iR is almost e_jR-injective. Suppose $m_n > 1$. Then the Kuppisch series can be started with e_jR. The condition reduces to the case $i < j$. Thus, either $m_i = m_j + j - i$ or $m_j = m_i + n + i - j$.

Suppose $m_n = 1$. Let $m_j \leqslant m_i$. We get a monomorphism $\eta : e_jR \to e_iR$. As e_iR has no two isomorphic composition factors, $\eta(e_jR) = e_iJ^{j-i}$, so that $m_i = m_j + j - i$. We now take $m_i < m_j$. Let $\sigma : A \to e_iR$, $A < e_jR$ be a maximal homomorphism that cannot be extended from e_jR to e_iR. Let $K = \ker\sigma$. Now $soc(\frac{e_jR}{K}) \cong soc(e_iR)$. As e_iR has no two composition factors isomorphic, $K = 0$. We get a monomorphism $h : e_iR \to e_jR$, which inverts σ on $\sigma(A)$. As e_jR has no two composition factors isomorphic, $h(e_iR) = e_jJ^{n+i-j}$,*i.e.* $A = e_jJ^{n+i-j}$. So that $m_j = m_i + n + i - j$.

(iii) Because of (i) and (ii), we need to prove only the converse. Let this condition hold. Suppose $m_i = m_j + (j - i)$. As $m_j < m_i$, e_iR is e_jR-injective. To prove e_jR is almost e_iR, we take a homomorphism $\sigma : A \to e_jR$, $A < e_iR$. Let $K = \ker\sigma$. If $K = 0$, we have $h : e_jR \to e_iR$, which inverts σ on $\sigma(A)$. Suppose $K \neq 0$. Then $soc(\frac{e_iR}{K}) \cong soc(e_jR) \cong soc(e_iR)$. Thus $d(K) = nv$ for some $v > 0$. Then $m_i - d(K) = m_j + j - i - nv < m_j$, therefore σ can extended be from e_iR to e_jR. This proves that e_jR is e_iR.

Suppose $m_j = m_i + n + i - j$. Then $m_n > 1$ and the Kuppisch series of R can be started with e_jR, which reduces this case to the earlier one. □

LEMMA 3.5. *Let $i < j$ and $soc(e_iR) \ncong soc(e_jR)$. Then the following hold.*

(i) There exists no partial homomorphism from e_iR into e_jR if and only if $m_i + i - j < m_j \leq n + i - j$.

(ii) There exists no partial homomorphism from e_jR into e_iR, if and only if $m_j - n + j - i < m_i \leq j - i$.

Proof. (i) Suppose there exists no partial homomorphism from e_iR into e_jR. In this case e_jR is e_iR-injective. If a composition factor of e_iR is isomorphic to $soc(e_jR)$, we get a non-zero partial homomorphism from e_iR into e_jR, which contradicts our hypothesis. Hence $m_i < n$. Suppose $m_j > n + i - j$. Then, as $\frac{e_jJ^{n+i-j}}{e_iJ^{n+i-j+1}} \cong \frac{e_iR}{e_iJ}$, we get a non-zero homomorphism from e_iR into e_jR, which is a contradiction. Hence $m_j \leq n + i - j$. But $m_i \leq m_j + j - i$, we get $m_i + i - j \leq m_j \leq n + i - j$.

Conversely, let $m_i + i - j < m_j \leq n + i - j$. Suppose we have a partial homomorphism $\sigma : L \to e_jR$, $L \leqslant e_iR$. As $m_i < m_j$, σ extends to a homomorphism $\eta : e_iR \to e_jR$. Then $\eta(e_iR) \subseteq e_jJ^{n+i-j}$. However, $e_jJ^{n+i-j} = 0$, as $m_j \leq n+i-j$.

We get $\eta(e_iR) = 0$, which is a contradiction. This proves (i). Similarly, we can prove (ii). □

LEMMA 3.6. *For some $i < j$, let $soc(e_iR) \not\cong soc(e_jR)$.*

(i) Let there exist a partial homomorphism from e_iR into e_jR and A be smallest length submodule of e_iR such that $soc(\frac{e_iR}{A}) \cong soc(e_jR)$. Then e_jR is almost e_iR-injective if and only if $d(A) = m_i - m_j + n + i - j$.

(ii) Let there exist a partial homomorphism from e_jR into e_iR and B be smallest length submodule of e_jR such that $soc(\frac{e_jR}{B}) \cong soc(e_iR)$. Then e_iR is almost e_jR-injective if and only if $d(B) = m_j - m_i + j - i$.

(iii) Let there exist a partial homomorphism from e_iR into e_jR and a partial homomorphism from e_jR into e_iR. If e_iR and e_jR are relatively almost injective, then $d(A) + d(B) = n$.

Proof. (i) Now there exists a partial homomorphism from e_iR into e_jR. So, we can get a smallest length $A < e_iR$ such that $soc(\frac{e_iR}{A}) \cong soc(e_jR)$. Then $A \neq 0$ and $d(A) < n$. Suppose e_jR is e_iR-injective. As e_jR is essentially e_iR-injective, we get a monomorphism from $\frac{e_iR}{A}$ into e_jR, which gives $m_i - d(A) \leq m_j$. For any homomorphism $\sigma : e_iR \to e_jR$, $\sigma(e_iR) \subseteq e_jJ^{n+i-j}$ and also there exists a homomorphism η from e_iR onto e_jJ^{n+i-j}. Therefore $\frac{e_iR}{A} \cong e_jJ^{n+i-j}$. Thus $m_i - d(A) = m_j - (n + i - j)$, $d(A) = m_i - m_j + n + i - j$. Conversely, let $d(A) = m_i - m_j + n + i - j$. Then $m_i - d(A) = m_j - (n + i - j)$, $m_i - d(A) \leq m_j$. As any uniserial module C_R is injective relative to a uniserial module D_R, whenever $d(D) \leq d(C)$, it follows that e_jR is e_iR-injective. Similarly, (ii) holds. (iii) is immediate from (i) and (ii). □

We now return to the main result Theorem (3.3). Its part (i) concerns that case when $soc(e_iR) \cong soc(e_jR)$ and is covered by (3.4). Now suppose $soc(e_iR) \not\cong soc(e_jR)$. Part (ii) concerns the case, when there is no partial homomorphism from e_iR into e_jR and no partial homomorphism from e_jR into e_iR and is covered by (3.4). Part (iii) concerns the case, when there exists a partial homomorphism from e_iR into e_jR, but there exists no partial homomorphism from e_jR into e_iR and is covered by (3.6)(i) and (3.5)(ii). The other part (iv) follows from (3.6)(ii) and (3.5)(i). Finally part(v) follows from (3.6).

Acknowledgement

The author is extremely thankful to the referee for his suggestions.

References

[1] F. W. Anderson and K. R. Fuller, *Rings and categories of modules*, Graduate Texts in Mathematics, vol. 13, Springer-Verlag, New York-Heidelberg, 1974. MR0417223

[2] A. Alahmadi and S. K. Jain, *A note on almost injective modules*, Math. J. Okayama Univ. **51** (2009), 101–109. MR2482408

[3] Y. Baba, *Note on almost M-injectives*, Osaka J. Math. **26** (1989), no. 3, 687–698. MR1021440

[4] C. Faith, *Algebra. II: Ring theory*, Springer-Verlag, Berlin-New York, 1976. MR0427349

[5] M. Harada, *Direct sums of almost relative injective modules*, Osaka J. Math. **28** (1991), no. 3, 751–758. MR1144484

[6] M. Harada, *Almost projective modules*, J. Algebra **159** (1993), no. 1, 150–157, DOI 10.1006/jabr.1993.1151. MR1231208

[7] M. Harada and A. Tozaki, *Almost M-projectives and Nakayama rings*, J. Algebra **122** (1989), no. 2, 447–474, DOI 10.1016/0021-8693(89)90229-9. MR999086

[8] S. Singh, *Almost relative injective modules*, Osaka J. Math. **53** (2016), no. 2, 425–438. MR3492807

[9] S. Singh, *Uniform almost relative injective modules*, J. Algebra **478** (2017), 353–366, DOI 10.1016/j.jalgebra.2017.01.033. MR3621678

[10] R. B. Warfield Jr., *Serial rings and finitely presented modules*, J. Algebra **37** (1975), no. 2, 187–222, DOI 10.1016/0021-8693(75)90074-5. MR0401836

House No. 424, Sector 35A, Chandigarh-160022, India
Email address: ossinghpal@yahoo.co.in

Selected Published Titles in This Series

738 **Shrikrishna G. Dani, Surender K. Jain, Jugal K. Verma, and Meenakshi P. Wasadikar, Editors,** Contributions in Algebra and Algebraic Geometry, 2019

735 **Yanir A. Rubinstein and Bernard Shiffman, Editors,** Advances in Complex Geometry, 2019

734 **Peter Kuchment and Evgeny Semenov, Editors,** Differential Equations, Mathematical Physics, and Applications, 2019

733 **Peter Kuchment and Evgeny Semenov, Editors,** Functional Analysis and Geometry, 2019

732 **Samuele Anni, Jay Jorgenson, Lejla Smajlović, and Lynne Walling, Editors,** Automorphic Forms and Related Topics, 2019

731 **Robert G. Niemeyer, Erin P. J. Pearse, John A. Rock, and Tony Samuel, Editors,** Horizons of Fractal Geometry and Complex Dimensions, 2019

730 **Alberto Facchini, Lorna Gregory, Sonia L'Innocente, and Marcus Tressl, Editors,** Model Theory of Modules, Algebras and Categories, 2019

729 **Daniel G. Davis, Hans-Werner Henn, J. F. Jardine, Mark W. Johnson, and Charles Rezk, Editors,** Homotopy Theory: Tools and Applications, 2019

728 **Nicolás Andruskiewitsch and Dmitri Nikshych, Editors,** Tensor Categories and Hopf Algebras, 2019

727 **André Leroy, Christian Lomp, Sergio López-Permouth, and Frédérique Oggier, Editors,** Rings, Modules and Codes, 2019

726 **Eugene Plotkin, Editor,** Groups, Algebras and Identities, 2019

725 **Shijun Zheng, Marius Beceanu, Jerry Bona, Geng Chen, Tuoc Van Phan, and Avy Soffer, Editors,** Nonlinear Dispersive Waves and Fluids, 2019

724 **Lubjana Beshaj and Tony Shaska, Editors,** Algebraic Curves and Their Applications, 2019

723 **Donatella Danielli, Arshak Petrosyan, and Camelia A. Pop, Editors,** New Developments in the Analysis of Nonlocal Operators, 2019

722 **Yves Aubry, Everett W. Howe, and Christophe Ritzenthaler, Editors,** Arithmetic Geometry: Computation and Applications, 2019

721 **Petr Vojtěchovský, Murray R. Bremner, J. Scott Carter, Anthony B. Evans, John Huerta, Michael K. Kinyon, G. Eric Moorhouse, and Jonathan D. H. Smith, Editors,** Nonassociative Mathematics and its Applications, 2019

720 **Alexandre Girouard, Editor,** Spectral Theory and Applications, 2018

719 **Florian Sobieczky, Editor,** Unimodularity in Randomly Generated Graphs, 2018

718 **David Ayala, Daniel S. Freed, and Ryan E. Grady, Editors,** Topology and Quantum Theory in Interaction, 2018

717 **Federico Bonetto, David Borthwick, Evans Harrell, and Michael Loss, Editors,** Mathematical Problems in Quantum Physics, 2018

716 **Alex Martsinkovsky, Kiyoshi Igusa, and Gordana Todorov, Editors,** Surveys in Representation Theory of Algebras, 2018

715 **Sergio R. López-Permouth, Jae Keol Park, S. Tariq Rizvi, and Cosmin S. Roman, Editors,** Advances in Rings and Modules, 2018

714 **Jens Gerlach Christensen, Susanna Dann, and Matthew Dawson, Editors,** Representation Theory and Harmonic Analysis on Symmetric Spaces, 2018

713 **Naihuan Jing and Kailash C. Misra, Editors,** Representations of Lie Algebras, Quantum Groups and Related Topics, 2018

712 **Nero Budur, Tommaso de Fernex, Roi Docampo, and Kevin Tucker, Editors,** Local and Global Methods in Algebraic Geometry, 2018

For a complete list of titles in this series, visit the
AMS Bookstore at **www.ams.org/bookstore/conmseries/**.